D1087446

FUNDAMENTALS OF LABORATORY SAFETY

PHYSICAL HAZARDS IN THE ACADEMIC LABORATORY

FUNDAMENTALS OF LABORATORY SAFETY

PHYSICAL HAZARDS IN THE ACADEMIC LABORATORY

WILLIAM J. MAHN

VAN NOSTRAND REINHOLD
New York

Library of Congress Catalog Number 90-48591
ISBN 0-442-00166-5

Printed in the United States of America

Van Nostrand Reinhold
115 Fifth Avenue
New York, New York 10003

Chapman & Hall
2-6 Boundary Row
London SE1 8HN, England

Thomas Nelson Australia
102 Dodds Street
South Melbourne, Victoria 3205, Australia

Nelson Canada
1120 Birchmount Road
Scarborough, Ontario M1K 5G4, Canada

16 15 14 13 12 11 10 9 8 7 6 5 4 3 2 1

Library of Congress Cataloging in Publication Data

Mahn, William J.
Fundamentals of laboratory safety: physical hazards in the academic laboratory/William J. Mahn.
p. cm.
Includes index.
ISBN 0-442-00166-5
1. Chemical laboratories—Safety measures. 2. Chemicals—Safety measures. I. Title.
QD51.M293 1990
642′.028′9—dc20 90-48591
CIP

To my wife, Carol Wickell,
and
N. Irving Sax, mentor and good friend

CONTENTS

PREFACE

Fundamentals of Laboratory Safety: Physical Hazards in the Academic Laboratory contains the information most often needed by laboratory workers. It is designed to be the first-choice reference for those who require information on the general hazards in an academic laboratory. Intended for use in universities, colleges, high schools, and middle schools, the book is also suitable for individuals handling laboratory equipment and hazardous chemicals in engineering, research and development, clinical, quality control, and food science laboratories. In general, the book is an attempt to promote and improve safety in the laboratory.

Besides identifying and providing advice on how to eliminate the hazards commonly encountered in science labs, engineering labs, and even school technical shops (e.g., welding, woodworking, and metalworking), the chapters contained here discuss the general concepts of laboratory safety, safety inspections, rules of proper laboratory behavior, glassware hazards, electrical hazards, equipment hazards, biological hazards, compressed gas and cryogenic materials hazards, radiation hazards, noise hazards, laboratory ventilation, protective and emergency equipment, spills and fires, and first aid. The appendices offer information on toxicity and fire hazard rating scales, National Fire Protection (NFPA) labels, Department of Transportation (DOT) warning labels, a list of carcinogens and other hazardous laboratory substances, and the addresses and telephone numbers for NIOSH, OSHA, and EPA offices around the country.

This book and a companion volume, *Academic Laboratory Chemical Hazards Guidebook,* allow the laboratory worker (students, technicians, instructors, and scientists alike) to quickly find information on specific laboratory hazards and their management by use of the extensive table of contents and index. Both books are beneficial to workers not only in an academic science laboratory (both physical and biological), but also those in the engineering lab and even a machine shop. All these areas involve the use of hazardous materials and equipment. The goal has been to provide a volume that is more likely to contain the required information than any other single source. If the needed information is not in this book or its companion title, I have tried to direct the reader to the next best reference. I would like to thank Gene Falken for providing the spark to begin this project.

William J. Mahn

FUNDAMENTALS OF LABORATORY SAFETY

PHYSICAL HAZARDS IN THE ACADEMIC LABORATORY

1

GENERAL CONCEPTS OF LABORATORY SAFETY

LABORATORY SAFETY PROBLEM

Safety in the laboratory requires the same kind of continuing attention and effort common to research, teaching, and analytical techniques. The use of new or different techniques, chemicals, and equipment requires careful study, instruction, and supervision. It also may require consultation with individuals having special knowledge or experience.

It should not be assumed that students, technical staff, or support personnel have adequate knowledge about laboratory safety. The information explosion makes it difficult to keep up to date on the possible consequences of exposures to laboratory chemicals and the precautions needed to control the hazards of laboratory operations. Academic training or past job experience may not have exposed one to the hazards of many of the new techniques and materials coming into use in every discipline, and we know that safety is not given adequate time or emphasis in an academic curriculum.

Too little time in laboratory courses is devoted to teaching the general principles of safety. A survey of chemistry and chemical engineering professors indicates that few include toxicology in their courses. Most believe that instruction on that subject should begin in high school and continue through college and graduate school. Generally, students are expected to acquire toxicological knowledge on their own. However, even if we began teaching students about chemical hazards we would still have a problem with those that have been trained already.

The central problems in any effort to provide general safety, chemical hazard, and toxicity information to those who need it in a laboratory setting seem to be the absence of an organized, functioning, and available repository for such information, and a general disinclination on the part of many to report hazards and toxicity information.

TYPES OF LABORATORY HAZARDS

In the laboratory, there are many sources of potential hazards. There are "routine" dangers: broken glass, knives and cutting tools, foreign bodies in the eye, falls, back injuries or hernias from improper lifting, and electrical shock. Also present are the unique hazards of corrosive, flammable, and toxic chemicals, radioactive substances, and runaway chemical reactions.

The major hazards in the laboratory arise from the following sources:

Dangerous equipment
Toxic chemicals
Flammable reagents
Explosive materials
Radioactive substances
Compressed gases
Cryogenic gases

Laboratory equipment can cause burns, electrical shock, cuts, fires, and explosions.

Other health hazards arise from the toxic materials used routinely in an analytical laboratory. Toxic substances may exhibit acute or immediate toxicity, examples being cyanides, hydrogen sulfide, arsenic compounds, and iodine. Toxic materials may also exhibit chronic or long-term toxicity, examples being mercury, formaldehyde, and chromium (VI) compounds.

Flammable materials in the laboratory include many organic solvents such as methyl ethyl ketone and acetone. Many materials are explosive under the right conditions; acetylene, ethyl ether, and hydrogen are examples.

Radioactive substances are often found in analytical laboratories where they are used as sources for detectors. An example is the tritium source in electron capture gas chromatographs and x-ray absorption detectors. Radioactive tracers are very convenient in following extractions and other processes.

Compressed gases are extensively used in an analytical laboratory in conjunction with chromatographic equipment, various fields of spectrophotometry (especially atomic absorption), polarographic techniques, and other analytical methods. Cryogenic gases are also employed in superconducting experiments and some specialized analytical applications such as nuclear magnetic resonance spectroscopy.

All of the aforementioned items are required for a laboratory to perform its functions. Their hazards cannot be eliminated, only controlled and minimized.

Health and Safety Hazard Rating Scales

The health and safety hazard system described here is similar to and compatible with the hazard identification system of the National Fire Prevention Association (NFPA) in its *Hazardous Chemical Data Manual.* The system is also described in the National Institute of Occupational Safety and Health (NIOSH) manual, *Manual of Safety and Health Hazards in the School Science Laboratory.* The rating scale for both health and safety hazards uses numerals from zero to four. A rating of zero (0) for an experiment indicates relatively little hazard involved, whereas a rating of four (4) indicates a serious hazard that is potentially lethal. Health ratings are based on the toxicity or pathogenicity of materials, chemicals, and organisms; safety ratings are based on the flammability and reactivity of chemicals and materials, the nature of procedures, and the kind and nature of glassware, equipment, instruments, and small tools involved. Experiments receive overall ratings after consideration of the quantity of materials used, their mode of dispensing, and the matrix (conditions and methods of use) in which the hazardous materials exist. The following criteria are used to determine the ratings:

- Toxicity of chemicals
- NFPA health hazard criteria
- Flammability of materials used in the experiments
- NFPA reactivity of materials
- Pathogenicity of biological organisms
- Electrical hazard potential
- Hazards from small tools and cutting instruments
- Procedures involved in the experiments
- Hazards in manipulating laboratory instruments and glassware

Health Ratings Based on Toxicity. When possible, a quantitative basis for the index ratings is used. Occupational Safety and Health Administration (OSHA) time-weighted average exposure data and, when available, rat oral LD_{50} or LC_{50} data are used in compiling the ratings. When this information is not available, judgment is based on other pertinent toxicological information. The bases of the health ratings are described in Table 1–1.

In all microbiology experiments, the presence of pathogens is assumed. The hazard level is assigned based on the nature of the pathogenic effects of the organism.

Safety Ratings Based on Physical Hazards. For safety ratings, a quantitative basis is more difficult to use. If the experiment uses materials rated by NFPA, then that code is used as a basis. Again, the quantity, conditions, and sample matrix are considered in arriving at a rating number, and a zero (0) to four (4) scale is used for materials, procedures, equipment, and experiments.

A rating of 4 is assigned if very dangerous materials such as those rated NFPA Code 4 are used or if very dangerous equipment is used. Effectively, this includes substances with flashpoints below 23°C, (73°F) and boiling points under 38°C (100°F); materials that form explosive mixtures with air; equipment readily capable of fatal electrical shock; equipment with exposed high-voltage circuits; and power cutting tools that are unguarded.

A rating of 3 is assigned to experiments involving materials with flashpoints between 23°C (73°F) and 38°C (100°F); materials capable of explosion but requiring a strong initiating source or needing to be heated; and equipment capable of shock, burning, and cutting with severe results.

A rating of 2 is assigned to experiments using materials that must be heated above ambient temperature before they can ignite [e.g., liquids with a flashpoint of

TABLE 1–1. Health Hazard Rating Scale

Health Hazard Rating	LD_{50}	LC_{50}
4	<50 mg/kg or carcinogen	<100 ppm
3	50–500 mg/kg	100–1000 ppm
2	500–5000 mg/kg	1000–10,000 ppm
1	>5 g/kg	>10,000 ppm

Source: Manual of Safety and Health Hazards in the School Science Laboratory, NIOSH, Cincinnati, Ohio, 1980, p. 2.

38°C (100°F) or over, but not exceeding 93°C (200°F)]; reagents that can react violently but not detonate; and power tools that are insulated and cannot cut.

A rating of 1 is assigned when experiments involve materials that must be preheated before ignition can occur. Examples include materials that must be heated to 816°C (1500°F) for 5 min before ignition can occur and liquids with flashpoints above 93°C (200°F); materials that, although usually stable, can become unstable at elevated temperature; and hand tools capable of puncturing or cutting.

A rating of 0 is assigned when there is little possibility of injury from the equipment or procedures involved. Materials used in these experiments will not burn and are not violently reactive with water or fire exposure. No dangerous equipment is used.

A summary of the scale is contained in Table 1–2. Table 1–3 gives examples of health and safety ratings for some instructional chemistry lab experiments.

TABLE 1–3. Hazards Index for Example Instructional Chemistry Experiments

Class of Experiments	Health/Safety
Acids and bases	2/1
Chemical families	3/2
Chemical and physical change	2/2
Conductivity and ionization	2/1
Crystals	0/0
Density measurements	0/0
Equilibrium	1/1
Gas laws	4/1
Heat of reaction	2/3
Measurement	1/0
Organic chemistry	2/3
Oxidation and reduction	2/1
Qualitative analysis	0/1
Radiation chemistry	0/0
Scientific processes and procedures	0/1
Stoichiometry	2/2
Reaction rates	0/0
Solubility	2/1
Thermal chemical measurements	2/2

Source: Manual of Safety and Health Hazards in the School Science Laboratory, NIOSH, Cincinnati, Ohio, 1980, p. 5.

Chemical Hazards

Laboratories that use chemicals have diverse goals: research, product and process development, routine analysis or education. The use and distribution of chemicals must be consistent with the goals of the laboratory. Teaching laboratories and research laboratories have different goals and, therefore, different operating conditions from other types of labs. Although the work done in teaching laboratories can usually be designed to use chemical substances that have well-known properties, work in research laboratories frequently involves chemical substances that have completely unknown properties. Teaching laboratories often involve large numbers of relatively inexperienced students, whereas research laboratories usually involve a small number of experienced investigators assisted by technicians.

Chemical hazards may be divided into six categories: flammability, instability, reactivity, corrosivity, toxicity, and radioactivity.

The risk associated with the possession and use of each specific substance is dependent on the following:

1. The knowledge and commitment to safe laboratory practices of all who handle it.

TABLE 1–2. Hazards Rating Scale

Scale	Health	Safety
4	Brief exposure could result in death.	Serious accident 80–100% probable. Death or serious injury very likely to result from the accident.
3	Prolonged exposure could result in death. Brief exposure could result in serious injury.	Accident 60–80% probable. Death or serious injury likely to result from the accident.
2	Prolonged exposure could result in serious injury. Brief exposure could result in mild injury.	Accident 40–60% probable. Injury may result from the accident.
1	Prolonged exposure could result in irritation or a mild injury.	Accident 20–40% probable. Slight injury may result from the accident.
0	Prolonged exposure should not result in irritation or injury.	Accident 0–20% probable. No injury should result from the accident.

Source: Manual of Safety and Health Hazards in the School Science Laboratory, NIOSH, Cincinnati, Ohio, 1980, p. 6.

2. Its physical, chemical, and biological properties.
3. The quantity received and the manner in which it is stored and distributed.
4. The manner in which it is used.
5. The manner of disposal of the substance and its derivatives.
6. The length of time it is on the premises.
7. The number of persons who work in the area and have open access to the substance.

The decision to procure a specific quantity of a specific chemical is a commitment to handle it responsibly from receipt to ultimate disposal. Each operation in which it is handled and each period between operations present opportunities for misadventure.

Have on hand and maintain up-to-date toxicological and chemical hazard data for all chemicals to be used in the laboratory, if such information is available. This information should be readily accessible to all concerned; for example, emergency personnel, safety officers, and firefighters.

The processes involved in the procurement, use, and disposal of chemicals are summarized in *Academic Laboratory Chemical Hazards Guidebook*, Chaps. 3–5. The safety coordinator and representative group must monitor these processes and make provisions for the orderly disposal of the material should circumstances such as spills or accumulation of unusable or hazardous material dictate such action.

The undeniable hazard of handling a variety of chemicals and legal requirements provide good and sufficient reasons for laboratories to bring their operations into compliance with current practice.

Handling Chemicals in the Laboratory. Chemicals occur in almost limitless (and ever-increasing) varieties. For this reason, general precautions for handling almost all chemicals are needed, rather than specific guidelines for each chemical. Otherwise, laboratory work will be needlessly handicapped, practically and economically, by attempts to adhere to a labyrinth of separate guidelines or, more likely, the laboratory worker will simply ignore the entire complex set of guidelines and, consequently, be exposed to excessive hazard.

Under the proper conditions, most chemicals can be hazardous. This book and *Academic Laboratory Chemical Hazards Guidebook* recommends chemical handling procedures that are aimed at minimizing all chemical hazards in the laboratory. In addition, special precautions for working with substances known to be flammable, explosive, or unusually toxic are described.

It is imperative that the work occurring in teaching and experimental research laboratories be differentiated from that in pilot plants and industrial manufacturing. Research in academic and industrial laboratories is carried out on a small scale and, hence, generally involves low levels of exposure of laboratory workers to chemicals. This is particularly true when the laboratory worker makes proper use of the hoods, protective apparel, and other safety devices that should be present in a well-equipped laboratory. Furthermore, in contrast to the typical industrial plant, where workers may be exposed to a limited number of substances over very long periods, the laboratory worker is exposed to a large variety of substances at low levels for brief periods of time. Finally, the professional expertise, common sense, judgment, and safety awareness of the worker performing chemical operations in the laboratory most often put him or her in the best position to judge necessary safety precautions. The problem is that all too often, that judgment is not made responsibly.

Careful attention must be paid to the appropriateness of the experimental work conducted in relation to the adequacy of the physical facilities available and the personnel involved. Once these points are established, it is the role of the safety coordinator and representative group to assist in the development of adequate guidelines for operations. For example, the ventilation facilities in a given laboratory may preclude certain kinds of work or the use of certain materials.

A continuing appraisal of safety facilities (hoods, incinerators, etc.) should be made, and modernization should be instituted whenever such facilities are judged inadequate for the work planned.

Disposal of Waste Substances. Some laboratories currently dispose of waste substances by pouring them down the drain or by placing them in drums to be buried in a landfill by an outside contractor. Such indiscriminate disposal is unacceptable and is being curtailed by a combination of local, state, and federal regulations. It is important that an institutional safety plan provide for the regular disposal of waste chemicals. Waste from individual laboratories should be removed at intervals of not more than 1 week to a central waste disposal storage area and then removed from that area at regular intervals. The most practical alternative for removal of combustible materials is to construct or contract for access to an incinerator that is capable of incinerating chemical and biological waste materials in an environmentally acceptable manner. The institutional plan for this type of disposal must include consideration of what materials can be incinerated, how they are to be collected and stored, and their mode of transport to the incinerator. See *Academic Laboratory Chemical Hazards Guidebook*, Chap. 5 for information on chemical disposal.

GENERAL RECOMMENDATIONS FOR SAFE PRACTICES IN LABORATORIES

It is impossible to design a set of rules that will cover all possible hazards and occurrences. Some general guidelines are given below that experience has shown to be useful for avoiding accidents or reducing injuries in the laboratory.

The most important rule is that everyone involved in laboratory operations, from a person at the highest administrative level to the individual worker, must be safety-minded. Safety awareness will become a habit on the part of everyone only if the issue of safety is discussed repeatedly and only if senior and responsible staff evince a sincere and continuing interest and are perceived as such by all their associates. The individual, however, must accept responsibility for carrying out his or her own work in accordance with good safety practices and should be prepared in advance for possible accidents by knowing what emergency aids are available and how they are to be used.

The supervisor of the laboratory has overall safety responsibility and should provide for regular formal safety and housekeeping inspections (at least quarterly for universities and other organizations that have frequent personnel changes and semiannually for other laboratories), in addition to continual informal inspections. Laboratory supervisors are responsible for ensuring that (1) workers know safety rules and follow them, (2) adequate emergency equipment in proper working order is available, (3) training in the use of emergency equipment has been provided, (4) information on special or unusual hazards in nonroutine work has been distributed to laboratory workers, and (5) an appropriate safety orientation has been given to individuals when they are first assigned to a laboratory space. The laboratory worker should develop good personal safety habits: (1) eye protection should be worn at all times, (2) exposure to chemicals should be kept to a minimum, and (3) smoking and eating should be avoided in areas where chemicals are present.

Advance planning is one of the best ways to avoid serious incidents. Before performing any chemical operation, the laboratory worker should consider the possibility of certain accidents or occurrences and be prepared to take proper emergency actions.

Overfamiliarity with a particular laboratory operation may result in a worker's overlooking or underrating its hazards. This attitude can lead to a false sense of security, which frequently results in carelessness. Every laboratory worker has a basic responsibility to himself or herself and colleagues to plan and execute laboratory operations in a safe manner.

General Principles

Every laboratory worker should observe the following rules:

1. Know the safety rules and procedures that apply to the work being done. Determine the potential hazards (e.g., physical, chemical, biological) and appropriate safety precautions before beginning any new operation.
2. Know the location of and how to use the emergency equipment in your area, as well as how to obtain additional help in an emergency, and be familiar with emergency procedures.
3. Know the types of protective equipment available and use the proper type for each job.
4. Be alert to unsafe conditions and actions and call attention to them so that corrections can be made as soon as possible. Someone else's accident can be as dangerous to you as any you might have.
5. Avoid consuming food or beverages or smoking in areas where chemicals are being used or stored.
6. Avoid hazards to the environment by following accepted waste disposal procedures. Chemical reactions may require traps or scrubbing devices to prevent the escape of toxic substances.
7. Be certain all chemicals are correctly and clearly labeled. Post warning signs when unusual hazards, such as radiation, laser operations, flammable materials, biological hazards, or other special problems exist.
8. Remain out of the area of a fire or personal injury unless it is your responsibility to help meet the emergency. Curious bystanders interfere with rescue and emergency personnel and endanger themselves.
9. Avoid distracting or startling any other worker. Practical jokes or horseplay cannot be tolerated at any time.
10. Use equipment only for its designed purpose.
11. Position and clamp reaction apparatus thoughtfully in order to permit manipulation

without the need to move the apparatus until the entire reaction is completed. Combine reagents in appropriate order and avoid adding solids to hot liquids.

12. Think, act, and encourage safety until it becomes a habit.

Health and Hygiene

1. Wear appropriate eye protection at all times.
2. Use protective apparel, including face shields, gloves, and other special clothing or footwear as needed.
3. Confine long hair and loose clothing when in the laboratory.
4. Do not use mouth suction to pipet chemicals or to start a siphon; a pipet bulb or an aspirator should be used to provide vacuum.
5. Avoid exposure to gases, vapors, and aerosols. Use appropriate safety equipment whenever such exposure is likely.
6. Wash well before leaving the laboratory area. However, avoid the use of solvents for washing the skin. They remove the natural protective oils from the skin and can cause irritation and inflammation. In some cases, washing with a solvent might facilitate absorption of a toxic chemical.

Food Handling

Contamination of food, drink, and smoking materials is a potential route for exposure to toxic substances. Food should be stored, handled, and consumed in an area free of hazardous substances.

1. Well-defined areas should be established for the storage and consumption of food and beverages. No food should be stored or consumed outside of this area.
2. Areas where food is permitted should be prominently marked and a warning sign (e.g., EATING AREA—NO CHEMICALS) posted. No chemicals or chemical equipment should be allowed in such areas.
3. Consumption of food or beverages and smoking should not be permitted in areas where laboratory operations are being carried out.
4. Glassware or utensils that have been used for laboratory operations should never be used to prepare or consume food or beverages. Laboratory refrigerators, ice chests, cold rooms, and such should not be used for food storage; separate equipment should be available for that use and prominently labeled.

Housekeeping

There is a definite relationship between safety performance and orderliness in the laboratory. When housekeeping standards fall, safety performance inevitably deteriorates. The work area should be kept clean, and chemicals and equipment should be properly labeled and stored.

1. Work areas should be kept clean and free from obstructions. Cleanup should follow the completion of any operation or at the end of each day.
2. Wastes should be deposited in appropriate receptacles.
3. Spilled chemicals should be cleaned up immediately and disposed of properly. Disposal procedures should be established and all laboratory personnel informed of them; the effects of other laboratory accidents should also be cleaned up promptly.
4. Unlabeled containers and chemical wastes should be disposed of promptly, by appropriate procedures. Such materials, as well as chemicals that are no longer needed, should not accumulate in the laboratory.
5. Floors should be cleaned regularly; accumulated dust, chromatography adsorbents, and other assorted chemicals pose respiratory hazards.
6. Stairways and hallways should not be used as storage areas.
7. Access to exits, emergency equipment, controls, and such should never be blocked.
8. Equipment and chemicals should be stored properly; clutter should be kept to a minimum.

Equipment Maintenance

Good equipment maintenance is important for safe, efficient operations. Equipment should be inspected and maintained regularly. Servicing schedules will depend on both the possibilities and consequences of failure. Maintenance plans should include a procedure to ensure that a device out of service cannot be restarted.

Guarding for Safety. All mechanical equipment should be adequately furnished with guards that prevent access to electrical connections or moving parts (such as the belts and pulleys of a vacuum pump). Each laboratory worker should inspect equipment before using it to ensure that the guards are in place and functioning.

The careful design of guards is vital. An ineffective guard can be worse than none at all, because it can give a false sense of security. Emergency shutoff devices may be needed, in addition to electrical and mechanical guarding.

Shielding for Safety. Safety shielding should be used for any operation with explosion potential, such as (1) whenever a reaction is attempted for the first time (small quantities of reactants should be used to minimize hazards), (2) whenever a familiar reaction is carried out on a larger than usual scale (e.g., 5–10 times more material), and (3) whenever operations are carried out under nonambient conditions. Shields must be placed so that all personnel in that area are protected from hazard.

Glassware

Accidents involving glassware are a leading cause of laboratory injuries.

1. Careful handling and storage procedures should be used to avoid damaging glassware. Damaged items should be discarded or repaired.
2. Adequate hand protection should be used when inserting glass tubing into rubber stoppers or corks or when placing rubber tubing on glass hose connections. Tubing should be fire-polished or rounded and lubricated, and hands should be held close together to limit the movement of glass should fracture occur. The use of plastic or metal connectors should be considered.
3. Glass-blowing operations should not be attempted unless proper annealing facilities are available.
4. Vacuum-jacketed glass apparatus should be handled with extreme care to prevent implosions. Equipment such as Dewar flasks should be taped or shielded. Only glassware designed for vacuum work should be used for that purpose.
5. Hand protection should be used when picking up broken glass. (Small pieces should be swept up with a brush into a dustpan.)
6. Proper instruction should be provided in the use of glass equipment designed for specialized tasks, which can represent unusual risks for the first-time user. (For example, separatory funnels containing volatile solvents can develop considerable pressure during use.)

See Chap. 3 for more information.

Flammability Hazards

Because flammable materials are widely used in laboratory operations, the following rules should be observed:

1. Do not use an open flame to heat a flammable liquid or to carry out a distillation under reduced pressure.
2. Use an open flame only when necessary and extinguish it when it is no longer actually needed.
3. Before lighting a flame, remove all flammable substances from the immediate area. Check all containers of flammable materials in the area to ensure that they are tightly closed.
4. Notify other occupants of the laboratory in advance of lighting a flame.
5. Store flammable materials properly.
6. When volatile flammable materials may be present, use only nonsparking electrical equipment.

See Chap. 12 for more information.

Cold Traps and Cryogenic Hazards

The primary hazard of cryogenic materials is their extreme coldness. They, and surfaces they cool, can cause severe burns if allowed to contact the skin. Gloves and a face shield may be needed when preparing or using some cold baths.

Neither liquid nitrogen nor liquid air should be used to cool a flammable mixture in the presence of air because oxygen can condense from the air, which leads to an explosion hazard. Appropriate dry gloves should be used when handling dry ice, which should be added slowly to the liquid portion of the cooling bath to avoid foaming over. Workers should avoid lowering their heads into a dry ice chest: Carbon dioxide is heavier than air, and suffocation can result. See Chap. 8 for more information.

Systems Under Pressure

Reactions should never be carried out in, nor heat applied to, an apparatus that is a closed system unless it is designed and tested to withstand pressure. Pressurized apparatus should have an appropriate relief device. If the reaction cannot be opened directly to the air, an inert gas purge and bubbler system should be used to avoid pressure buildup. See Chap. 8 for more information.

Waste Disposal Procedures

Laboratory management has the responsibility for establishing waste disposal procedures for routine and emergency situations and communicating these to laboratory workers. Workers should follow these procedures with care, to avoid any safety hazards or damage to the environment. See *Academic Laboratory Chemical Hazards Guidebook*, Chap. 5 for more information.

Warning Signs and Labels

Laboratory areas that have special or unusual hazards should be posted with warning signs. Standard signs and symbols have been established for a number of special situations, such as radioactivity hazards, biological hazards, fire hazards, and laser operations. Other signs should be posted to show the locations of safety showers, eyewash stations, exits, and fire extinguishers. Extinguishers should be labeled to show the type of fire for which they are intended. Waste containers should be labeled for the type of waste that can be safely deposited in them.

The safety- and hazard-sign systems in the laboratory should enable a person unfamiliar with the usual routine of the laboratory to escape in an emergency (or help combat it, if appropriate).

When possible, labels on containers of chemicals should contain information on the hazards associated with use of the chemical. Unlabeled bottles of chemicals should not be opened; such materials should be disposed of promptly and will require special handling procedures. See *Academic Laboratory Chemical Hazards Guidebook*, Chap. 4 for more information.

School laboratories should make liberal use of available signs and symbols in order to promote a safer environment. There are few mandated signs or symbols. Display of the radiation symbol is required by federal and state law wherever radioactive materials are being used.

Biohazard signs may also be required by state or federal law. Fire extinguisher signs are required by state and local law. All effective laboratory operations utilize signs and symbols as an important element of their safety program.

Stock wording is available from many vendors in signs made of enameled sheet metal or laminated fiber-glass. Self-adhering cloth or plastic signs are also available with standard wording.

Unattended Operations

Frequently, laboratory operations are carried out continuously or overnight. It is essential to plan for interruptions in utility services such as electricity, water, and inert gas. Operations should be designed to be safe, and plans should be made to avoid hazards in case of failure. Whenever possible, arrangements for routine inspection of the operation should be made and, in all cases, the laboratory lights should be left on and an appropriate sign placed on the door.

One particular hazard frequently encountered is the failure of cooling water supplies. A variety of commercial or homemade devices can be used that (1) automatically regulate water pressure to avoid surges that might rupture the water lines or (2) monitor the water flow so that its failure will automatically turn off electrical connections and water supply valves.

Working Alone

Generally, it is prudent to avoid working alone in a laboratory building. Under normal working conditions, arrangements should be made between individuals working in separate laboratories outside of working hours to cross-check periodically. Alternatively, security guards may be asked to check on a laboratory worker. Experiments known to be hazardous should not be undertaken by a worker who is alone in a laboratory.

Under unusual conditions, special rules may be necessary. The supervisor of a laboratory has responsibility for determining whether certain work requires special safety precautions, such as having two persons in the same room during a particular operation.

Accident Reporting

Emergency telephone numbers to be called in the event of a fire, accident, flood, or hazardous chemical spill should be posted prominently in each laboratory. In addition, the home numbers of the laboratory workers and their supervisors should be posted. These persons should be notified immediately in the event of an accident or emergency. Adverse reactions from laboratory exposures to chemicals and physical and biological agents should be reported.

Every laboratory should have an internal accident-reporting system to help discover and correct unexpected hazards. This system should include provisions for investigating the causes of injury and any potentially serious incident that does not result in injury. The goal of such investigations should be to make recommendations to improve safety, not to assign blame for an accident. Relevant federal, state, and local regulations may require particular reporting procedures for accidents or injuries.

Everyday Hazards

Finally, laboratory workers should remember that injuries can and do occur outside the laboratory or other work areas. It is important that safety be practiced in offices, stairways, corridors, and other places. Here, safety is largely a matter of common sense, but a constant safety awareness of everyday hazards is vital.

Environment

Chemicals must be disposed of in such a way that people, other living organisms, and the environment generally are subjected to minimal harm by the substances used or produced in the laboratory. Both laboratory workers and supporting personnel should know and use acceptable disposal methods for various chemicals.

SAFETY PLANS

Organizations administering laboratories should have safety plans. The goals of a laboratory safety plan should be to protect from injury those working in the laboratory, others who may be exposed to hazards from the laboratory, and the environment. The administration should actively support safety by ensuring that a safety plan is developed and followed. All persons in the organization must understand their responsibilities and should take appropriate actions to ensure safe operations. In many organizations, safety committees representing the various types of persons exposed to potential hazards are needed to discuss problems, recommend safety measures, and facilitate communication. Most organizations should have safety coordinators to work on problems, to serve as consultants on safety matters, and to support those involved in providing for safe operations. Safety programs should be a regular, continuing effort and not merely a stand-by activity that functions for a short time after each laboratory accident.

Administrative program planning is normally done by the "higher ups" in many organizations, including schools. However, safety program planning is at least one great exception to this rule. In industry, management and supervision do not provide a safe workplace without employee participation. In schools, few if any administrators and department heads can have safe laboratories without teacher and student cooperation.

The laboratory safety guidelines and rules to be used within a single organization should be developed with the active participation of the groups that will be affected, to assure better acceptance and to obtain the best possible ideas. If for any reason safety rules are developed unilaterally at the administrative level, it is strongly recommended that the rules be distributed as "proposed," with the time and opportunity for laboratory personnel to ask questions and register objections. Such a delay will result in better rules with less need for exceptions. A safety plan should include the following.

Facilities

Available facilities are an important part of the provision for safe laboratory operations; their capacity should not be exceeded. The facilities should include hoods, an appropriate ventilation system, stockrooms and storerooms, safety equipment, and arrangements for the disposal of materials. The performance of the laboratory ventilation system and other safety facilities should be monitored at regular intervals, at least once every three months. In cases where the facilities are inadequate for the work being done, they should be improved so that they are adequate or the experimental work should be changed so that the safety capacity of the facilities is not exceeded.

Although the energy costs of ventilation, often substantial, are increasing, considerations of economy should never take precedence over ensuring that laboratories have adequate ventilation. Any change in the overall ventilation system to conserve energy should be instituted only after the thorough testing of its effects has demonstrated that the laboratory workers will continue to have adequate protection from hazardous concentrations of airborne toxic substances. An inadequate ventilation system can cause increased risk because it can give a false sense of security. See Chap. 9 for more information.

Academic Teaching Laboratories. In general, the students and instructors in academic teaching laboratories should follow the safety procedures recommended for full-time laboratory workers in research and development laboratories. The need for using appropriate protective apparel (safety glasses, gloves, etc.), for following general safe laboratory practices, and for providing emergency safety equipment (safety showers, eyewash stations, fire extinguishers, etc.) is probably even greater in instructional laboratories where sizable numbers of relatively untrained laboratory workers may be present in relatively close quarters.

The use of safety equipment by students and instructors in science laboratories is an integral part of a laboratory safety program. Safety equipment is a broad designation that includes exhaust hoods, personal protective devices such as respirators and safety eyeglasses, and fire extinguishers.

The most severe limitation on protective equipment in instructional laboratories is usually the general laboratory ventilation and, especially, auxiliary local exhaust ventilation (hoods or their equivalent). It is unlikely that most academic institutions will be able to provide a laboratory hood for every two students in their instructional laboratories. The work done and the chemicals used in any instructional laboratory should be adjusted according to the quality of ventilation protection that is available in that laboratory. Unless adequate hood space can be provided, it seems prudent to avoid work with substances whose toxicity has not been studied. The selection of particular substances whose toxicological properties are known should be based on the quality of the ventilation system available.

The permissible exposure limits (PELs of OSHA) and current threshold limit values (TLVs; see the hazardous chemical entries, *Academic Laboratory Chemical Hazards Guidebook*) provide useful guides. All work in instructional laboratories should be carried out in such a way that the concentration of each substance being used does not exceed its PEL or recommended TLV. This may be achieved by a combination of experimental design and laboratory ventilation. In general, the use of a hood or some equivalent form of local ventilation is desirable when working with any appreciably volatile substance having a TLV of less than 50 ppm. Of course, this generalization is intended to serve as no more than a rough rule of thumb; many substances having higher limit values can pose hazards if they are used without proper planning and precautions. Furthermore, the overall ventilation system in each instructional laboratory should be evaluated at regular intervals, and some monitoring of concentration levels for specific substances may be required in questionable cases.

The Uniform Fire Code requires 20 ft^2 of laboratory space per student in academic science laboratories. This is an area about 4.5 ft on a side. The same code requires 50 ft^2 in vocational education labs. This recommendation is expected to include storage and preparation areas that serve all students. California's Administrative Code [Title 2, Subchapter 4, Section 1811 (g)] requires 1300 ft^2 for 24 students (54 ft^2 per student), which includes preparation and chemical storage space.

The National Science Teacher's Association (NSTA) has recommended for safety considerations that science classes should be limited to 24 students unless a team of teachers is available.

Toxic substances (health hazard rating of 4) should not be used in instructional laboratories unless there is a fume hood for every two students.

Monitoring of Chemical Substances

For most laboratory environments, the regular monitoring of the airborne concentrations of a variety of different toxic materials is both unjustified and impractical. If care is taken to ensure that (1) the ventilation system (including the hood) is performing and being used properly, (2) the laboratory workers are wearing proper protective clothing, and (3) the laboratory workers are following good hygiene and laboratory safety practices, then even highly toxic materials can be handled without undue hazard.

There are three circumstances in which the monitoring of individual compounds is appropriate:

1. If a specific substance that is highly toxic is regularly and continuously used (e.g., three times a week), then instrumental monitoring of that substance may be appropriate. This is especially true if a relatively large amount of the material is being stored or used in the laboratory.
2. Radioactive substances should always be monitored during all phases of work.
3. In testing or redesigning the hoods and other local ventilation devices in a laboratory, it is often helpful to release a substance (e.g., Freon 11^{RT} or sulfur hexafluoride) whose airborne concentration is readily monitored by commercial instruments. Alternatively, laboratory workers can wear personal air-sampling devices to provide a measure of the airborne concentrations of some substance in their environment.

Medical Program

Any person whose work involves the regular and frequent handling of toxicologically significant quantities of material that is acutely or chronically toxic should consult a qualified physician to determine whether it is desirable to establish a regular schedule of medical surveillance. It can be very useful to monitor body concentrations of chronic toxins such as lead or mercury compounds.

The need for regular medical surveillance should be decided on an individual basis by consultation between the laboratory worker and a qualified physician. Copies of medical records should be retained by the institution in accord with state and federal regulations. Local, state, or federal regulations sometimes require medical surveillance for specific compounds.

Safety Training

Training is an important part of a safety program. Safety training is a critical responsibility of the instructor and his or her immediate supervisor. Supervisors should periodically check the results of safety training and review or repeat as necessary.

Each individual working in a laboratory should receive training about safety in connection with his or her laboratory and its ongoing work. This group includes faculty, research staff, students, laboratory supervisors, laboratory workers, maintenance and storeroom personnel, and others who might be physically close to laboratories. New persons coming into the laboratory or employees functioning in related jobs should be educated about safety procedures and the appropriate actions to take in the event of accidents.

Stockroom personnel, maintenance personnel, technical assistants, animal care personnel, persons transporting chemicals, and others in the vicinity of the laboratory may also be exposed to potential physical and chemical hazards in connection with the laboratory's ongoing work. They should be informed about the risks involved and educated about how to avoid potential hazards and what to do in the event of an accident.

These institutional education programs should be regular, continuing activities and not simply once-a-year presentations provided for groups of new students or employees.

Sources of Safety Information

Literature and consulting advice on laboratory safety and on the physical and biological hazards of chemicals should be readily available to those responsible for laboratory operations and to those directly involved. Laboratory workers should be encouraged to read about the potential hazards of their laboratory's ongoing work and to learn about the availability of various resources describing safe operating conditions. This literature should be available in a form that is readily accessible both to those responsible for laboratory operations and to laboratory workers themselves.

Although a substantial number of people with expertise in laboratory safety are employed by large chemical companies and private consulting firms, such persons are not often found in academic institutions. Because modifications of certain safety facilities (e.g., ventilation systems and waste disposal systems) can be very expensive, such modifications should not be undertaken until advice has been sought from persons qualified to make recommendations. The alternatives for an academic institution are either to hire an appropriately qualified person as the institutional safety coordinator or to hire appropriate consultants as needed to obtain advice about specific safety problems. Some chemical companies have discussed the possibility of encouraging contact between their staff safety experts and universities to provide information about safety problems that may arise in a particular academic setting. Such interactions would have immediate benefit for universities and could be expected to increase the safety consciousness of students being trained there. Similar programs could be designed for colleges and secondary schools.

Material Safety Data Sheets (MSDS). Material Safety Data Sheets should be available for all chemicals in the laboratory. They should be kept in an organized file so that all laboratory workers can find them easily.

ORGANIZING FOR SAFETY

Laboratory safety requires a continuous, comprehensive, and coordinated effort to be effective. Piecemeal attempts on a voluntary basis will simply not do the job. Everybody must "get into the act" as a regular routine for safety programs to work.

Safety planning should be integrated directly into the experimental procedure, and the reasons for actions taken should be understood. To conduct experimental work in the laboratory without consideration of the hazards involved and the potential environmental impact of such work is tantamount to science in the dark ages. NIOSH, OSHA, and EPA are facts of life, and students deserve to be exposed to these government agencies as part of their academic scientific training. The regulatory aspects of these agencies are binding on us all and will influence our way of life for years to come.

A sound safety organization that is respected by all is essential; a good laboratory safety program must always be based on the participation of both the laboratory administration and students and/or employees. Laboratory workers and their institutions or companies are strongly encouraged to follow recommended safety practices. Large industrial organizations often have safety programs costing millions of dollars annually. Academic institutions and small companies have been rather casual about safety programs; the time and money devoted to safety programs by these organizations will have to increase. University faculties and administrations need to develop, support, and enforce well-defined safety policies. The importance of a safety-minded point of view among all employees or students must be inculcated by the institution. In the end, the individual worker must learn to consider possible hazards and seek information and advice before beginning any experiment.

A successful safety program must recognize potential hazards before an accident occurs.

Safety Policy

The basic approach of laboratory safety is a safety policy that is followed by all workers and visitors. The safety policy must be known and understood by each instructor and student. A manual with rules should be adopted and made available to all. This policy should also stress that everyone is responsible for his or her own safety, as well as that of co-workers.

Safety Committee

A safety committee of representative workers can be organized to provide input, but it should not be made responsible for safe laboratory conditions and practices. A revolving safety inspection team is a good organizational approach, since more judgment is brought to the problem. However, the ultimate responsibility resides with management.

Responsibility for Laboratory Safety

First and foremost, the protection of health and safety is a moral obligation. An expanding array of federal, state, and local laws and regulations makes it a legal requirement and an economic necessity as well. In the final analysis, laboratory safety can be achieved only by the exercise of sound judgment by informed, responsible individuals. It is an essential part of the development of scientists that they learn to work with and accept responsibility for the appropriate use of hazardous substances.

For an organization, responsibility for safety within its laboratories may be considered to exist at three different levels: organizational or institutional, supervisory or instructional, and individual. The division of responsibility, in this or any other appropriate manner, needs to be clearly assigned and accepted, steps taken to see that the responsibilities are exercised, and the division reassessed if unexpected problems develop.

Liability for a laboratory misadventure (accident, illness, environment damage) may lie with the individual experimenters, their immediate supervisors, other officers of the institution, or the institution itself, depending on the circumstances and applicable laws—federal, state, and local. In view of the small number of cases decided, it is impossible to predict the outcome of a lawsuit in this area. Each institution, therefore, should seek expert legal advice pertinent to its particular situation, so that potential liability can be estimated ahead of time if possible.

The ultimate responsibility for safety within an institution lies with its chief executive officer. This individual must ensure that an effective institutional safety program is in place. The chief executive officer and all immediate associates (deans, department heads, etc.) must exhibit a sincere and continuing interest in the safety program, and this interest must be obvious to all. An excellent safety program that is ignored by top management (until after an accident) will certainly be ignored by everyone else.

The organization or institution of which the laboratories are part has a fundamental responsibility to provide the facilities, equipment, and maintenance necessary for a safe working environment, or an organized program to make the improvements required for a safe working environment. Unless the institution or organization fulfills its responsibilities, it cannot expect its supervisors, employees, or students to fulfill their responsibilities for laboratory safety.

An institutional safety coordinator (or officer) is essential to an effective institutional safety program. This individual should have appropriate training and be qualified in those areas of safety relevant to the activities of the institution. Records should document that the facilities available and the precautions taken in carrying out activities of the institution are compatible with current knowledge of potential risks and the law. Experimental work involving chemicals is a subset of those activities.

The responsibility for safety in a department (or other administrative unit) lies with its chairperson or supervisor. Usually, this responsibility is delegated to a departmental safety coordinator (or officer). In smaller institutions, it may be possible for one person to perform more than one set of duties. For example, often a departmental faculty member might also serve as a departmental safety coordinator. However, it must be recognized that such duties are time-consuming and will require regular attention.

To be effective, a departmental safety coordinator must be committed to the attainment of a high level of safety and must work with administrators and investigators to develop and implement policies and practices appropriate for safe laboratory work. In these activities, the safety coordinator requires the cooperation of everyone—workers, technicians, and students, as well as scientists. Collectively, this group must routinely monitor current operations and practices, see that appropriate audits are maintained, and seek ways to improve the safety program. If the goals of the laboratory dictate specific operations and the use of specific substances not appropriate to the existing facilities, it is the responsibility of the safety coordinator and the representative group to assist the investigator in acquiring adequate facilities and developing appropriate guidelines.

The instructor or supervisor has the responsibility of giving all necessary directions, including the safety measures to be used, and seeing that students or employees carry out their individual responsibilities. Whoever directs the activities of others has a concurrent responsibility to prevent accidental injuries from occurring as a result of these activities. The responsibility for safety during the execution of an operation lies with the operator(s) executing that operation. Operators frequently include workers, technicians, and students. Nevertheless, the primary responsibility remains with the instructor or supervisor.

Each individual who works in a laboratory as a student or employee has a responsibility to learn the health and safety hazards of the chemicals he or she will be using or producing, and the hazards that may occur from the equipment and techniques employed, so that he or she may design a setup and procedures to limit the effects of any accident. The individual should investigate any accident that occurs, and record and report the apparent causes and preventative measures that may be needed to prevent similar accidents.

In some academic institutions and companies with laboratories, there seems to be a general lack of responsibility in some important areas concerning safety. Whether this neglect is due to a lack of adverse experience or a general tendency toward freedom from supervision, the potential for accidents in laboratories calls for sophisticated attention to measures for controlling hazards and limiting injury frequency and severity.

Past experience has shown that voluntary safety programs are often inadequate. Good laboratory practice requires mandatory safety rules and programs. To achieve safe conditions for the laboratory worker, a program must include the following and assign responsibilities for their implementation:

1. Preparation and practice of fire, emergency, and rescue procedures.
2. Formal and regular safety programs ensuring that at least some of the full-time personnel are trained in the proper use of emergency equipment and procedures.
3. Training in cardiopulmonary resuscitation.
4. Regular safety inspections at intervals of no more than 3 months (and at shorter intervals for certain types of equipment, such as eyewash fountains). See Chap. 2 and 10.
5. Labeling containers of chemicals.
6. Providing collection and disposal procedures at regular intervals for waste chemicals and other hazardous wastes. See *Academic Laboratory Chemical Hazards Guidebook*, Chap. 5.
7. Providing consultative assistance in laboratory safety and occupational health.
8. Establishing realistic policies on working alone.
9. Requiring eye protection.
10. Providing proper laboratory ventilation and regularly monitoring the operation of this equipment. See Chap. 9.
11. Evaluating toxic exposures through biospecimen analysis by medical personnel.
12. Providing appropriate immunization and protection against biohazards.

As these responsibilities are more broadly and fully exercised, the frequency, severity, and cost of laboratory accidents and injuries will decline.

2

SAFETY INSPECTIONS AND RECORDKEEPING

INSPECTIONS

In this section, we are going to consider safety inspections, operations of safety audits, and information surveys.

One of the most common safety activities in the laboratory is the safety inspection. It can be very effective if done well, or it can be useless if a cursory inspection is made by an inadequately trained individual or group.

Of concern to all laboratory personnel is what happens after such an inspection, what happens to any report on it, and what action is taken as a result of the effort made in conducting the inspection. If no improvement is made, the inspection might as well have not been made.

Safety Professional as an Inspector

The safety officer or safety professional should be constantly making inspections. Whenever he is in a hazardous area, he should be noting problems, unsafe conditions, unsafe acts, etc. At some fixed interval, he should perform a formal safety inspection. To do this in a thorough, effective way takes a great deal of time and effort. Checklists or inspection sheets should not be relied on totally. They do have their use, but can prevent one from overlooking some critical aspect of the area being inspected. The inspector should not ascribe primary importance to the checklist and neglect unsafe items simply because they do not appear on it. A safety professional will ask questions about what is being done, what materials are being used, under what conditions the work is being conducted, what precautions are being taken, what investigation preceded the work, etc. To do this well, and to obtain the amount of information needed to make a good safety inspection, the inspector needs to be a good diplomat and also have an exceedingly keen eye.

General Guidelines for Conducting an Inspection

1. Review the previous period's inspection report and carry over uncompleted items on this report as indicated.
2. Concentrate most on those areas with the greatest accident exposure. Remember to inspect for off-the-floor conditions as well as those at ground level.
3. If one is available, study a checklist of what to look for. See Tables 2–1, 2–2, and 2–3.
4. Search for evidence of property damage as clues to unsafe behavior. Suggest controls to reduce damage and potential injury losses.
5. Place the date each item was discovered immediately after the item in parentheses. This information should be transferred to other reports when the item is not complete.
6. When you report on remedial action in your followup, make sure you indicate what intermediate safety measure is being taken when a permanent remedy will require some time.
7. Make sure you know how many copies of a report to make and to whom to send them. Print clearly or type. Make sure final reports are filed properly for record purposes.
8. Encourage supervisors to take such action in the correction of unsafe conditions so that a permanent remedy is instituted. Correcting the same item repetitively is costly.
9. Place an asterisk in front of any item you feel requires special fast attention because of hazard severity.
10. Be persistent and regular in your own followup.

TABLE 2–1. Walkthrough Survey of the School Science Laboratory

Complete the following pages for your school. If desired, send copies to your principal, superintendent, and state science supervisor. Be sure to keep one copy on file for yourself. Repeat at least annually.

Name ______________________ Date __________

Title ______________________

School ______________________

Department ______________________

Location: City ______________________ State __________

1. Number of science teachers in the department. __________
2. Number of science teachers with safety training. __________
3. Number of science laboratories. __________
4. (a) Recommended student capacity. __________
   (b) Actual student capacity. __________
5. Number of science classrooms. __________
6. Number of combined science classrooms/laboratories. __________
7. (a) Number of exits per laboratory. __________
   (b) Locate exits by layout diagram (attach). __________
   (c) Are exits properly marked? Yes ___ No ___
   (d) Are storage rooms properly marked? Yes ___ No ___
8. Number of fire extinguishers: __________

Type	*Number*	*Location*
(a) CO_2	______	______________________
(b) Soda acid	______	______________________
(c) BC	______	______________________
(d) ABC	______	______________________
(e) Water	______	______________________
(f) Halon	______	______________________

9. Number of backup fire extinguishers in storage. __________
10. Last inspection date for fire extinguishers. __________
11. Number of sand buckets with sand. __________
12. Number of approved fire blankets. __________
13. Number of spill cleanup kits. __________
14. List any special spill cleanup kits. __________
15. Number of first-aid or emergency charts. __________
16. Number of first-aid kits with supplies. __________
17. Number of safety showers. ______ Are they operable? __________
    (a) Industrial type. __________
    (b) Hand or portable type. __________

TABLE 2–1. Walkthrough Survey of the School Science Laboratory *(Continued)*

18. Number of eyewash stations. ____________
    (a) Installed with plumbing and aerifier. ____________
    (b) Squeeze-bottle type. ____________
    (c) Pressurized canister type. ____________
    (d) Other. ____________
19. Are they checked regularly to determine operability? Yes ___ No ___
20. Are they flushed regularly to minimize microbial contamination? ____________
21. Eye, face, and body protection. ____________
    (a) Number of approved safety glasses with full side shields. ____________
    (b) Number of approved safety chemical goggles. ____________
    (c) Number of approved plastic face shields. ____________
    (d) Number of demonstration safety shields. ____________
22. (a) Does each student have his or her own personal eye protection device? Yes ___ No ___
    (b) If the answer is no, is there a maintenance and cleaning program? Yes ___ No ___
23. Number of rubber gloves. ____________
24. Number of rubber/plastic/cloth aprons. ____________
25. Number of asbestos gloves (pairs). ____________
26. Number of lab coats. ____________
27. Number of electric outlets (110–120 V without ground). ____________
28. Number of electric outlets (110–120 V with ground). ____________
29. Number of electric outlets (110–120 V with ground fault interrupter). ____________
30. Has provision been made for proper grounding of all electrical devices? Yes ___ No ___
    If no, please describe.____
    ____
    ____
    ____
31. (a) Number of compressed gas cylinders. ____________
    (b) Are they properly secured to prevent tipping? Yes ___ No ___
32. Number of sinks. ____________
33. Number of waste receptacles for glass. ____________
34. Number of waste receptacles for dry chemicals or reagents. ____________
35. Number of waste receptacles for liquid chemicals or reagents. ____________
36. Number of containers designed to transport dangerous reagents or chemicals. ____________
37. Are all waste receptacles properly marked? ____________
38. Are all waste receptacles easily located? ____________
39. Number of securable storage spaces for chemicals with forced ventilation. ____________
40. Number of securable storage spaces for chemicals without forced ventilation. ____________

41. Number of electric refrigerators. ____________
42. Briefly describe what type of materials are stored in these units, such as reagents, food, etc.

____

____

43. (a) Number of fume hoods. ____________
    (b) Rated exhaust velocity—cfm, if known. ____________
44. (a) Number of exhaust fans. ____________
    (b) Rated exhaust velocity—cfm, if known. ____________
45. Are the fume hoods rated explosion-proof? ____________
46. Number of drinking fountains in science rooms? ____________
47. Number of drinking fountains in science labs? ____________
48. Is there a master cut-off for water? ____________
49. Is there a master cut-off for gas? ____________
50. Is there a master cut-off for electricity? ____________
51. (a) Are they accessible? ____________
    (b) Do you know where they are? ____________
52. Are the floors nonskid? ____________
53. Is there sharp-edged furniture in the lab? ____________
54. Number of special cabinets to store hazardous or flammable chemicals. ____________
55. (a) Number of gas burners. ____________
    (b) Number of alcohol burners. ____________
56. (a) Do you use animals in the lab? ____________
    (b) If so, are there proper facilities to handle them? ____________
57. Are you aware of the biohazards involved in animal handling? ____________
58. (a) Are experiments conducted using biologic fluids? ____________
    (b) What is the source? ____________
59. (a) Does blood-letting experimentation take place? ____________
    (b) If so, are disposable lancets and alcohol swabs used? ____________
60. Do you permit handling of pathogens by students? ____________

    If yes, explain. ____

____

____

____

61. Do you have pipet bulbs for proper pipetting procedures? ____________
62. Is food preparation/consumption/storage permitted in the laboratory? ____________
63. Do you have proper facilities to accommodate handicapped students? ____________

    Explain here. ____

____

____

____

TABLE 2-1. Walkthrough Survey of the School Science Laboratory *(Continued)*

64. Do you have field manuals or explanatory sessions describing the possible dangers of field trips?	______
65. Do you feel properly trained to conduct safe and healthful science laboratory experimentation in your school? If no, contact your state science supervisor concerning available training.	______
Other Comments	

Hazard Evaluation Equipment

A thorough safety inspection may require more than just the visual observation of a hazardous area. Chemicals and physical agents constantly pose a threat to safety and health where they are made, used, and stored. There are many testing instruments that can be used to evaluate hazards in the laboratory. A velometer can be used to measure face velocity in a fume hood. A radiation detector can be used to look for radioactive contamination. See Table 2–4 for other instruments that may be useful for hazard evaluation.

Measuring devices must be properly maintained and calibrated. Calibration should be constantly checked, but it is especially important when the devices are received, repaired, and in hard use. Other equipment maintenance procedures are the following:

1. Always store in a cool, dry, dust-free noncorrosive atmosphere.
2. Check battery-operated instruments for full charge before use.
3. Keep records of battery changes, calibration, and maintenance, plus any alterations in the equipment.
4. Study and follow the manufacturer's directions. File them in a convenient place and duplicate for lab use.

Committee Inspections

Many inspections are conducted by committees. A safety committee, newly appointed by management, will often decide on inspections as the best way to get started. A small committee can function well and, if it has a reasonable amount of training, can do quite a significant job. A large committee generally bogs down and achieves little.

Management Inspections

When a manager, particularly the director of the laboratory, takes the time and shows the interest to go on an inspection tour, the impact of his or her presence is very appreciable. Generally, only a part of the facility can be inspected in the short amount of time made available for such a purpose. However, if laboratory personnel do not know in advance where an inspection is going to take place, it will have a salutary effect on the housekeeping and general appearance of the entire laboratory. Management inspections are strongly recommended. The participation of the manager in this role serves to convince everyone that safety is indeed important.

Insurance Inspectors

Sometimes, insurance carriers will send engineers to conduct inspections of laboratory facilities. Generally, they will zero in on a few aspects, such as the handling of flammable liquids, or possibly the use of high-pressure equipment. Many times their recommendations are beneficial, and they certainly should be taken seriously. The safety officer, if one is available, should always accompany the insurance inspector, both to answer questions and to place the situation in proper perspective for the insurance inspector.

Special-Purpose Inspections

Some of the best inspections are made for special purposes—that is, giving attention to particular cate-

TABLE 2–2. Hazardous Chemicals Survey

Please answer yes, no, not applicable or give an appropriate answer as indicated.

A. Flammable Liquids

1. Are flammable solvents stored in your lab?__________
2. What volume (gal) of flammables do you keep in your lab?__________
3. What volume (gal) of flammables do you keep elsewhere?__________
4. Where are these flammables stored if not in the laboratory? (Be specific: for example, Bldg. 119, Rm. 202.)

   ____
5. Do you use OSHA-recommended flammable storage lockers for flammable storage?__________
6. What is the approximate volume of flammable solvents used in your lab on a daily basis?

   ____
7. Do you have in your lab(s)?

   (a) Fire extinguishers.__________

   (b) Safety showers.__________

   (c) Eyewash station. (capable of 15 min continuous stream).__________

   (d) Fire blankets.__________

   (e) Automatic fire extinguishers.__________

   (f) Posted exit instructions in case of a fire.__________

B. Corrosive Chemicals

1. What acids and bases are stored and used in your lab?

Acids *Volume Stored*	*Bases* *Volume Stored*
1. __________	1. __________
2. __________	2. __________
3. __________	3. __________
4. __________	4. __________
5. __________	5. __________
6. __________	6. __________
7. __________	7. __________
8. __________	8. __________

2. Are these stored in an appropriate acid or base storage locker?

   ____
3. Are they stored:

   (a) Together?__________

   (b) Near flammables?__________

   (c) Near toxic chemicals?__________

   (d) Near compressed gases?__________

   (e) Near cryogenic gases?__________

TABLE 2–2. Hazardous Chemicals Survey *(Continued)*

C. Toxic Chemicals

1. Does your lab use toxic chemicals?

   Yes? ________

   No? ________

   Not sure? ________

2. If yes, please list those chemicals you believe to be most toxic (see Appendix D and the hazardous chemicals entries in *Academic Laboratory Chemical Hazards Guidebook*).

   (a) ________

   (b) ________

   (c) ________

   (d) ________

   (e) ________

   (f) ________

   (g) ________

   (h) ________

   (i) ________

   (j) ________

3. Do you work with mercury in your lab? ________
4. Where do you store your toxic chemicals? ________
5. In your estimate, how old are most of the chemicals on your shelves?

   (a) New within the last year. ________

   (b) 1–2 years old. ________

   (c) 2–3 years old. ________

   (d) 3–5 years old. ________

   (e) More than 5 years old. ________

   (f) Don't know. ________

6. Do you work with any of the following chemicals in your lab?

   Peroxides. ________

   Chlorinated solvents. ________

   Ethers. ________

   Fulminates. ________

   Perchloric acid. ________

   Picric acid. ________

   Azides. ________

   Cyanides. ________

   Acetylene. ________

D. Compressed Gases and Cryogenic Liquids

Please list those compressed gases and cryogenic liquids used in your lab.

Compressed Gases	*Cryogenic Liquids*
1. ________	1. ________

2. ______	2. ______
3. ______	3. ______
4. ______	4. ______
5. ______	5. ______

E. Please list any radioactive materials used in your lab.

______

______

______

______

Source: Adapted from Mell, L.D., *Laboratory Safety Evaluation Survey,* Naval Medical Research and Development Command, Washington, D.C., 1980, pp. 26–28.

gories of hazards. (See Table 2–3 for an example hazardous chemical survey.) When two- or three-man committees are inspecting, each member of the committee can make one of the categories his priority, although he is not necessarily excluded from making observations in other fields.

One person might look for electrical hazards, noting any frayed or defective cords, seeing that covers are on junction boxes, and ensuring that other electrical equipment is in repair.

Another committee member might cover flammable liquids, observing whether excessive amounts exist in glass containers and whether safety cans are used.

Special-purpose inspections covering fire protection are of particular importance. The presence and condition of all fire extinguishers are noted, any extinguishers blocked by materials or equipment are made accessible, and any obstructions that would prevent quick exit from a hazardous area are noted and corrected.

Operations Safety Audit

The operations safety audit is a way of achieving an in-depth study of a process of operation, with a view to uncovering as many as possible of the weak points or potential hazards in the operation. It is particularly suited to pilot plants or similar complex functions.

Such a study results by building a team with people of various disciplines. For example, the team should include a process expert or a supervisor of the operation, a design engineer, a maintenance engineer, experts in such areas as electrical work, piping, etc., if these factors are involved; it should also include a safety professional who functions as secretary for the study.

This team should meet with adequate time to conduct a thorough, unhurried review of the entire operation in order to get an overview. They must also cover all points of stress, controls, wear, and maintenance requirements. Modifications that have been introduced, injuries that have occurred, and all significant breakdowns should be noted. The reasons for any undesirable occurrences should be ascertained, and plans should be made to alter or correct the operation to forestall future accidents, injuries, and damage to equipment.

A complete report should be drafted, with priorities given to changes to be made. This schedule of changes should be made a part of the report so that when the next study is conducted, all such information will be available for its audit committee.

RECORDKEEPING

Recordkeeping has always been an integral part of laboratory operations and also a normal part of the instructor's duties. Instructors are required to keep attendance and grade records. Laboratories keep records as part of good laboratory practice, and occasionally, as a result of a legal requirement to do so.

It is a common misconception that laboratories *must* keep records of analyses and daily activities. This is not true. Other than for legally regulated laboratories, for example, clinical and hospital laboratories, there is no legal requirement to maintain daily work records. Some voluntary standards organizations such as the American Society for Testing and Materials (ASTM) and the American Industrial Hygiene Association (AIHA) have established voluntary accreditation programs requiring that accredited laboratories maintain a detailed recordkeeping system.

TABLE 2–3. Survey of General Laboratory Operations and Practices

1. Is pipetting by mouth allowed (or practiced) in your lab?
   ____
2. Are any of the following permitted in the lab?
   (a) Eating. ______
   (b) Drinking. ______
   (c) Smoking. ______
   (d) Application of cosmetics. ______
3. How are used needles disposed?
   ____
   ____
4. Which of the following types of hoods are in your lab? (Please indicate the number of each.)
   (a) Fume hood. ______
   (b) Class I Biological Safety Cabinet. ______
   (c) Class II Biological Safety Cabinet (A or B). ______
   (d) Class III Biological Safety Cabinet. ______
   (e) Other. ______
   (f) Don't know. ______
5. How often are the above hoods (cabinets) checked for proper air flow or, for Class II cabinets, certified? ______
6. How are chemical wastes disposed?
   ____
   ____
   ____
7. Do you store chemicals:
   (a) Alphabetically ______
   (b) Or by special classes? ______
   (See *Academic Laboratory Chemical Hazards Guidebook,* Chap. 4.)
8. What procedures in your lab do you consider hazardous?
   ____
   ____
   ____
   ____
9. Is eye protection required for lab personnel who handle chemicals and/or biohazards?
   ____
10. Are contact lenses permitted in your lab? ______
11. Are safety goggles and/or safety face shields provided in your lab? ______
12. Are lab coats worn in your lab? ______
13. Are lab coats permitted to be worn outside your lab? ______
14. Do you consider your lab to be:
    (a) Nonhazardous? ______
    (b) Slightly hazardous? ______

(c) Moderately hazardous? ______

(d) Very hazardous? ______

(e) Extremely hazardous? ______

15. Are laboratory personnel familiar with emergency procedures for:

(a) Caustic burns? ______

(b) Electrical shock? ______

(c) Inhalation of toxic chemicals? ______

(d) Biological accidents that involve infectious agents? ______

(e) Radioactive spills? ______

16. Are laboratory staff members allowed to work alone? (If yes, indicate if they are allowed to work alone after normal duty hours.)

______

17. If workers are allowed to work alone, is there a policy for assuring the safety of such workers? ______

18. Do you work with lasers in your research? ______

Source: Adapted from Mell, L.D., *Laboratory Safety Evaluation Survey,* Naval Medical Research and Development Command, Washington, D.C., 1980, pp. 28–29.

However, there are no statutes, state or federal, which require that daily work records be kept.

Even though school science laboratories probably do not keep records that relate directly to laboratory operation as opposed to those that relate directly to students' performance, they should be encouraged to do so. Maintaining adequate records is a key element of any successful safety program. Good records will allow laboratory personnel, both teachers and students, to spot trouble before it occurs, or determine causes if an accident happens.

TABLE 2–4. Equipment for Evaluating Hazards

Item	Purpose
Dosimeter	Measure radiation exposure (dose) over a period of time. See Chap. 11.
Electrical multimeter	Evaluate static charge, current leaks, and electrical outlets.
Flammable gas detector	Take air measurements.
Geiger–Mueller counter	Measure levels of radioactivity. See Chap. 11.
Light meter	Measure levels of illumination.
Sound level meter	Measure noise levels. See Chap. 17.
Velometer	Measure air flow. See Chap. 9.
Smoke tube	Measure air flow direction and rate. See Chap. 9.

Laboratory records are useful as an aid in delivering testimony while serving as a witness in a legal proceeding. If accurate records are kept routinely, they may be admitted as evidence in a court of law, even though the maker of such records is unavailable or cannot recall the events from which their record was drawn.

These two benefits are the primary reasons for establishing the recordkeeping psychology in students. As a matter of good practice, commercial laboratories require their employees to keep daily work records. Students should be taught as early as possible the benefits and desirability of adequate recordkeeping.

Chemical reagent inventory records, repair and inspection records, injury records, and incident records are extremely valuable in establishing an ongoing, viable safety and health program. They can also be useful to the instructor as a means of minimizing his or her legal liability in the case of bodily injury or property damage.

In some cases, the school may have no choice in deciding whether or not it wants to keep records. Schools subject to the provisions of the Occupational Safety and Health Act of 1970 (OSHA) must keep injury and incident records. These schools include all private and parochial schools and all public schools in states enforcing the OSHA under an acceptable state plan.

Inventory Records

There is no excuse for a school science laboratory not having readily available inventory records covering the chemical reagents in stock, the existing instrumentation, and miscellaneous items such as glassware, fire extinguishers, and personal protective devices.

All of these records play an important part in the operation of a laboratory safety and health program. If students are involved in a recordkeeping system, they can develop and acquire work habits that will stand them in good stead through their professional years.

Academic Laboratory Chemical Hazards Guidebook, Appendix G contains a sample chemical reagent inventory record with storage recommendations and hazards for each chemical.

Instrument Inventory Records

Instrument inventory records should record the dates of purchase, routine maintenance checks, and all repairs. These records will serve as a quality-control check on the instruments, and they may aid the instructor in defending a liability suit. Maintaining instruments in safe operating condition is a normal function of a science teacher. Written records are acceptable, legal proof that this responsibility has been effectively carried out by the teacher.

Repair and Inspection Records

Repair and inspection records are another facet of laboratory operations. Written records contribute valuable information about the day-to-day problems of running a laboratory, but most important, adequate records are indicative of an instructor who is attuned to the social, legal, and economic currents of the times.

The general public is more aware of health and safety problems now than at any time previously in the history of this country. The public demands that schools provide a safe and healthful learning environment for young people. This is a social policy.

Repair and inspection records are economic essentials. They serve to pinpoint equipment subject to an undesirable number of breakdowns. Equipment that breaks down often requires more frequent repairs and excessive spending. Equipment subject to frequent breakdown may present a greater safety and/or health hazard than more durable equipment. Equipment placed on a regular maintenance schedule supported by written records is much less likely to become a laboratory health or safety hazard.

Inspection and repair records also can serve as valuable elements of a legal defense. If a student is injured while using a particular piece of equipment, it is possible that a teacher could be charged with negligence in maintaining that item of equipment. Records indicating regular inspection and equipment repair can be used to refute such a claim of negligence.

Inspection records covering routine, periodic inspections are another element of a successful safety program. For schools and school districts amenable to OSHA regulations, inspections may be required in order to ensure that the school facilities and operations are in compliance with the applicable regulations. Laboratory facilities may have to comply with the regulations governing the storage of flammable liquids, adequate electrical wiring, signs and labels, and exposure to hazardous chemical and physical agents. Inspection records also provide information on potential problem areas and allow the teacher to take remedial action before any problem occurs. Routine inspections may be conducted by students as a way of increasing their participation in the safety program aspect of the laboratory operation.

Injury Records

There are two accident and injury recordkeeping systems in common use in this country. These are the American National Standards Institute (ANSI) Z16.1 system and the OSHA system. OSHA has promulgated regulations that govern the recording and reporting of occupationally induced illnesses and injuries. The OSHA recordkeeping requirements do not extend to students unless the students involved are serving as employees of the school.

The ANSI system is also designed primarily for use in case of industrial accidents. It is a more detailed system than the OSHA system and is probably beyond the scope of most school operations. However, the concepts represented by these two systems are important and should be extended to student injuries whenever possible.

Accident and injury records can be of inestimable value to a teacher in any future litigation regarding student injury. If the school has adopted a uniform system of accident reporting, these accident records will be readily admitted as evidence in any court of law.

Incident Records

Incidents may be defined as those events and occurrences in science laboratories that are undesirable and may, but not necessarily do, result in personnel injury. Spills of caustics or acids, small fires, breaking of glass, and electrical shock would qualify as such incidents. Keeping records of these incidents will serve to alert the laboratory teacher to possible danger areas

TABLE 2–5. Items to Include in a Safety Checklist

Item	Number and/or Type		
	Classroom	Laboratory	Storeroom
Door(s)	______	______	______
Window(s)	______	______	______
Floor	______	______	______
Cabinet(s)	______	______	______
Shelving	______	______	______
Ventilation	______	______	______
Lighting	______	______	______
Water supplies	______	______	______
Waste drains	______	______	______
Fire extinguishers	______	______	______
Safety showers	______	______	______
Fire blankets	______	______	______
Eyewash stations	______	______	______
First-aid kits	______	______	______
Fume hoods	______	______	______
Spill cleanup kits	______	______	______
Student lab stations	______	______	______
Instructor preparation station	______	______	______
Instructor demonstration desk	______	______	______
Master water cut-off valve	______	______	______
Master electric cut-off switch	______	______	______
Master gas cut-off valve	______	______	______
Chemical reagents	______	______	______
Disposition of hazardous materials	______	______	______
Disposition of unlabeled materials	______	______	______
Disposition of unusable equipment	______	______	______
Glassware	______	______	______
Heating equipment	______	______	______

Comments:

____

____

____

incidental to a particular laboratory or a particular group of students.

Incident records may not prove to be conclusive in a single school setting. Ideally, data should be collected over district and state territories in order to provide statistically valid information concerning the frequency of occurrence of specific incidents.

Student Assignments and Agreements

Recordkeeping in the laboratory serves the dual purpose of assisting the science teacher in providing a safe place for his students to work and learn and in minimizing his legal liability at the same time. The two purposes are not mutually exclusive. If the teacher accomplishes the first goal, he must necessarily accomplish the other. One aspect of the two duties of supervision and instruction is providing adequate instruction and information to the students on particular assignments or experiments.

SAFETY SURVEYS AND CHECKLISTS

Table 2–5 is a laboratory safety checklist that may be used by teachers to:

1. Determine whether or not a safe environment exists.
2. Indicate possible areas of concern and danger.
3. Serve as a guide for the design of safe facilities.
4. Act as a monitoring device for periodic safety checks.
5. Act as a permanent record of an ongoing safety program.

3

GLASSWARE

INTRODUCTION

One of the most common materials of construction in the science laboratory is glass. The great majority of containers used in science laboratories are made of glass. Glass is an excellent material of construction because it is relatively inexpensive, highly resistant to chemical attack, easily cleaned, and noncontaminating. Glass has one major disadvantage that counterbalances all of its advantages in the science laboratory: It breaks very easily. When glass breaks, it produces extremely sharp edges and points that readily cut human tissue.

Glass is a complex mixture of sodium, calcium, and magnesium silicates. Other materials (e.g., boron, lead, or phosphates) are added in order to achieve variations in physical properties.

PROPERTIES

Physical Properties

Glass is easy to fabricate. Instead of a sharp melting point, it has a super-cooled liquid-softening range. It retains thermal and mechanical stresses unless carefully annealed.

Mechanical Properties

Glass has very poor elasticity. Mechanical shock produces cracks, "stars," chips, and in some cases residual stresses that later produce failure. Glassware assemblies must be strain-free. Use flexible linkages in support structures. Scratches produce localized strain that can be useful (e.g., cutting glass tubing by scratching with a file) or can cause unpredictable failures.

Thermal Properties

Lime (ordinary) glass has a relatively high coefficient of thermal expansion. The thermal coefficient of expansion is decreased by adding boron, or by substituting pure silica (SiO_2, quartz, or substances with various trade names, e.g., "Vicor"). Thus, boron glass (Pyrex) has less thermal expansion and quartz the least (near zero over ordinary ranges). Thermometers cannot be annealed after manufacture and therefore must be handled carefully. They may be "seasoned" by repeated heating over the range before calibration.

Chemical Reactivity

Glass is inert to most chemicals except fluorides and strong alkalies, including concentrated ammonium hydroxide. These reagents weaken the glass and destroy surface integrity. Glass strongly adsorbs many chemicals that may alter or interfere with trace chemical analysis and sampling—for example, soap films, nitrogen oxides, mercury ions, benzene, etc.

Ion-Exchange Properties

Glass can be used as a semipermeable membrane in ion-selective electrodes.

Optical Properties

Glasses can be highly transparent to the visible spectrum, but become opaque in ultraviolet (UV) and infrared (IR). Quartz glass, fused silica, has good transparency to most UV radiation.

GENERAL HAZARDS OF LABORATORY GLASSWARE

Most laboratory accidents involving injuries involve glassware failure that is caused by improper use. Glassware is fragile. Rules for normal handling can be categorized by types of hazard and summarized as follows.

Fracture Hazard

When glass is fractured, the razor sharp edges of the fragments present an extreme cutting hazard. Special tempered glass produces fragments that are much less sharp and seldom produce wounds.

Thermal Hazards

1. Avoid shock created by sudden temperature changes. Even Pyrex will crack under extreme shock. Use quartz if conditions require such abuse.
2. There is increased shock sensitivity from stresses retained on cooling after heating. These stresses may not be noticed unless the glass is observed under polarized light (e.g., tempered safety glass lenses). Use only annealed glassware when possible. However, even annealed glassware can fail when used at a high temperature and then cooled rapidly.
3. Thermometers are not annealed and thus retain thermal stresses that can result in fractures, causing injury and releasing mercury.
4. Improperly annealed safety glasses can fail without warning.
5. "Do it yourself" fabrication (glass-blowing) creates thermal stress hazards—for example, bending glass tubing without annealing in a reducing (smokey) flame. See the section on fabrication on p. 31.
6. Residual stresses from high-temperature chemical reactions can weaken the glass.
7. Glass cools slowly. Manipulate heated glass with caution to avoid burns.
8. Wear hand protection when handling hot glassware.
9. Use tongs for lifting hot beakers, flasks, bottles, etc. Use grips or long-handled tongs for manipulating hot evaporating dishes, crucibles, etc.
10. Never heat soft glass that is thick-walled (bottles, jars, etc.). The glass will break at the zone between the thick and thin walls.
11. Never heat pipets, volumetric flasks, and burettes. These items can change volume as a result of expansion and contraction. They can also fail.
12. Do not heat bottles, graduated cylinders, volumetric glassware, funnels, jars, droppers, watchglasses, desiccators, glass plates, and test tubes. They are often made of soft glass, particularly in older apparatus. If so, they have low resistance to thermal shock. Soft glass tubing is widely used because it can be bent in the heat of a laboratory burner.

GENERAL RULES FOR HANDLING GLASSWARE

Never use cracked, scratched, chipped, or "starred" glassware. The failure of glassware is almost always initiated at its surface. Localized impacts from sharp objects, scratches, and heating can all place extremely high stresses on a glass surface. Total failure can quickly result at one of these high-stress locations. Glassware that reveals surface nicks and scratches on close observation should be removed from service and either discarded or repaired.

Never force a rubber stopper off a piece of glassware. Slice the stopper parallel to the axis of the glassware and cut it off. A rubber stopper is inexpensive compared to the cost of replacing a thermometer or distilling head, not to mention the potential for injury. Always protect hands with leather gloves or several layers of cloth when inserting glass.

To Protect Against Mechanical Shock

1. Protect bottoms of flasks and beakers by cementing on rubber or plastic pads.
2. Coat flasks, especially wash bottles, with vinyl latex that is available commercially.
3. Limit the storage of flammable or hazardous liquids to 500-mL bottles.
4. Do not store together noncompatible chemicals—for example, acids and alkalies, oxidizing and reducing agents such as concentrated ammonium hydroxide and nitric acid in the same tray, concentrated nitric and methanol, etc. (See *Academic Laboratory Chemical Hazards Guidebook,* Chap. 2.)
5. Use a safety carrier when transporting hazardous chemicals, especially liquids.
6. Use retaining basins under chemical reactions.
7. Install shock-pads on the bottom of glassware that frequently is picked up and set down on bench tops (e.g., wash bottles).
8. Wear side-shield goggles for eye protection when working with glassware or carrying bottles.
9. Use plastic containers when possible, squeeze bottles, etc., but beware of plastic volumetric ware that is not dimensionally stable.

10. Store glassware in a manner that will avoid the rough contact which produces "stars," cracks, and chips.

To Protect Against Mechanical Strains

1. Use flexible connections in glassware assemblies.
2. Avoid the use of more clamps than absolutely necessary to support assemblies. Use "floating" suspensions and platforms.
3. Flat surfaces are less resistant to mechanical shock than curved or spherical surfaces where stress is partially dissipated.

To Protect Against Explosion and Implosion Hazards

1. All pressure reactions in glass are highly dangerous.
2. The potential for severe hazard exists in high-velocity fragments, slivers, and shards. They can cause eye injury and loss of blood from severed blood vessels.
3. In vacuum applications, as the curvature of a vacuum container decreases (the surface becomes flatter), the hazard greatly increases; large flasks must be completely enclosed.

8. Install transparent shields around equipment that is under vacuum or in which explosions, runaway reactions, boil-overs, etc. may endanger the operator and bystanders. Remember, there is a person on the other side of the bench.
9. Stripe glassware with tape or wrap it with a plastic or wire screen if it is to be used under vacuum or pressure. This reduces the possibility of flying glass fragments in case of implosion or explosion.

RULES FOR USE OF SPECIFIC TYPES OF LABORATORY GLASSWARE AND PROCEDURES

Tubing

It is very easy to break glass tubing or rods when inserting them into a hole bored into a rubber or cork stopper. Before any tubing is inserted, care must be taken to ensure that the hole in the stopper has been bored to the correct size. Glass tubing will not penetrate a stopper just because the stopper has a hole in it. The diameter of the hole should be slightly smaller than the diameter of the glass rod or tubing. If there is a large size disparity, excessive force may be necessary to force the glass piece through the stopper. When excessive force is used, trouble is likely to occur. All forces used to insert the tubing should be directed along the axis of the tubing. Glass tubing or rods should never be subjected to bending or flexing forces while being inserted into a stopper.

Always lubricate glass tubing when inserting it into rubber stoppers. This rule applies to tubing, rod, thermometers, tubes, condenser and flash fittings, and all other glassware. Glycerol or water is an appropriate lubricant. Wear leather or heavy cotton lined plastic gloves when inserting tubing or thermometers through rubber stoppers, and when engaging in any other similar activity.

When inserting glass tubing into rubber or plastic tubing, many of the precautions discussed above should be observed. The glass should be lubricated before insertion into the rubber or plastic tubing. The rubber tubing should be cut at an angle before the insertion of the glass piece. The angled cut allows the rubber to stretch more readily.

Nearly all glassware cutting in school science laboratories involves glass rods or tubing. The individual doing the cutting should be wearing leather gloves and eye protection. Do not try to cut through glass tubing with a file. It will break up at the point of pressure and injuries may result. Use a tubing cutter on tubing of about ½ in. or larger. File scratches often do not give smooth breaks.

To cut small-diameter glass tubing and rods, make a scratch on the tubing with a sharp triangular file. With a quick motion, then snap the tubing at the scratch. The rod or tube should be grasped in both hands, one on each side of the score mark, and the thumbs should be extended and placed against the glass tubing or rod opposite the score mark. The rod or tubing should be bent toward the body, thus putting the scored surface in tension. The rod or tubing should break cleanly at the score mark. The cut ends should be fire-polished to eliminate any sharp edges. The glass is not "cut," but crystallized by the file scratch.

Cutting glass tubing and rods is basically a very simple but sometimes dangerous procedure. Hands must be well protected if injuries are to be avoided.

When bending glass tubing, care must be exercised not to crimp the bend. The tubing is particularly weak at the crimp and more subject to failure.

Pipets

Handling liquids with pipets is probably the greatest hazard in the laboratory. In most cases, the use of the mouth to apply suction should be avoided. A special controllable pipet bulb is the safest approach. However, aqueous solutions of nontoxic materials can be pipetted by mouth, although it is generally not good laboratory practice to pipet anything by mouth. Be careful not to introduce air in the tip during withdrawal because the pipet may empty into the mouth. Solvents or solutes with appreciable vapor pressure should always be pipetted with a bulb.

Pipet tips and mouthpieces can be easily broken, resulting in severe cuts of the hand and fingers. Mechanical stress should not be applied to the stem. A grip with the thumb over the top of the pipet as a delivery control may cause the pipet to break in the hand. The use of the index finger as a control is safer.

Burettes

The main danger in the use of burettes arises during any filling operation. The liquid should be added carefully so it does not dribble down the arm or spatter into the face. Severe burns can occur when filling burettes at elevated heights and some reagents dribble down an arm. A small beaker and funnel minimize this hazard. Safety glasses should always be worn. Remember that alkalies cause more serious burns than dilute acids.

Burettes should not be heated. They can change volume as a result of expansion and contraction.

Cleanliness is critical with burettes and pipets. Care must be taken when cleaning any glassware with chemical solutions.

Beakers

Hold a beaker with one hand around the side. Larger beakers (>500 mL) should be held with one hand around the side and one hand underneath. Larger beakers should be heavy-duty (with thicker glass). Beakers are most commonly broken against a lab bench top. Be cautious when setting down a beaker on a bench top and preferably clear the surrounding area before picking it up. Hot beakers should be handled with beaker forceps or tongs. Place hot beakers on a ceramic-centered gauze pad.

Volumetric Flasks

The principle hazard in using volumetric flasks is the mechanical failure of the flask when shaking it. Both the neck and bulb of the flask must be supported when shaking to keep the neck from breaking.

Do not use volumetric flasks for the storage of solutions. They are relatively unstable with their narrow bases and are difficult to sample from through their long, narrow necks. Volumetric flasks make very expensive storage bottles.

Volumetric flasks should not be heated. They will change volume as a result of expansion and contraction and may fail, as many are not made of annealed glass.

Crucible and Evaporating Dishes

Porcelain equipment may be heated over a direct flame. This always creates potential for failure; consider the possibility of violent reactions or explosions. The use of crucible fusions in the preliminary treatment of samples is extremely hazardous. Nickel, stainless steel, or platinum crucibles should be used when needed. All these processes should be conducted in a hood.

Cleaning Glassware

Many laboratory injuries occur when cleaning glassware. In addition to the possibility of injury from broken glass, the possibility of injury from so-called "cleaning solutions" also exists.

Wear rubber gloves wherever manipulation permits when washing glassware. Many cuts occur during this seemingly mundane activity.

Most glassware can be effectively cleaned by using any one of a number of laboratory detergents available for that purpose. Glassware not contaminated with a hard-to-remove residue can be soaked in a detergent-water solution, brushed to get rid of strongly attached materials, rinsed with tap water and distilled water, and then air-dried. It may be necessary to rinse glassware cleaned in this manner with a dilute acid in order to remove all of the detergent from the surface of the glass.

All chemical cleaners other than detergents are hazardous. The chemicals commonly used for this purpose are dichromate-acid, trisodium phosphate, and alcoholic potassium hydroxide. If totally contaminant-free glassware is required, acid cleaning of the glassware is commonly used to remove all trace contaminants from the glass surface.

Dichromate-acid cleaners are mixtures of potassium dichromate and concentrated sulfuric acid. This combination dissolves and oxidizes organic grease and scum. It also dehydrates, burns, and oxidizes skin, causing blisters or second-degree burns on contact.

Glassware to be acid-cleaned is first washed with a detergent solution, as was discussed above. The glassware is then soaked in a solution of sodium dichromate and sulfuric acid. The soaking period may vary depending on the exact purpose for which the glassware will be used. After soaking is completed, the glassware is rinsed with tap water and distilled water, and then allowed to air-dry.

Anyone following this procedure must wear the full complement of required personal protection: safety goggles, face shield, rubber gloves, rubber apron, and lab coat. She must use extreme care to avoid splashing herself or others with concentrated acids. She must not discharge excessive quantities of concentrated acid into a sink drain without taking proper precautions. Any acid flushed down a drain should be diluted with copious quantities of water.

Other acids, concentrated nitric and concentrated hydrochloric, can be used to replace or supplement the sodium dichromate-sulfuric acid cleaner when necessary. The instructor must realize that the acid-cleaning procedure can be extremely dangerous to students and any building fixtures. The chromic acid cleaning solution may also present a disposal problem.

Trisodium phosphate cleaners are strong alkalies. They may burn as well as irritate skin.

Alcoholic potassium hydroxide is prepared from a concentrated solution of potassium hydroxide that is diluted with alcohol. It causes burns and is also extremely flammable.

These cleaning solutions are dangerous and should be handled accordingly. A full disclosure of the hazards involved should be included on the label of the storage container.

There are some surfactant detergents available that may be used as a substitute for acid cleaning. The hazards associated with acid or strong alkali cleaners are severe and attempts at substitution should be made.

Vacuum-Related Hazards

The analyst often uses vacuum filtration, desiccation, and distillation. Filtering flasks, desiccators, and other such items should be stripped with tape to contain the glass in case of implosion. Glass containers should be designed for vacuum use with heavy walls and freedom from flaws. Plastic or wire screen should be used with taping. The vacuum pump pulley should be shielded. Safety glasses and other personal protective equipment must also be used.

Extraction Procedures

The separation procedure or extraction often utilizes potentially hazardous equipment and materials. Separation funnels may break at the stem. Extra care is necessary when dealing with toxic and flammable solvents. Appropriate personal safety equipment must be used. These procedures should be carried out in hoods if dangerous solvents are being used.

Do not attempt to extract a solution until it is colder than the boiling point of the extracting solvent. Do not point the funnel drain toward anyone or toward a flame. When a volatile solvent is used, the unstoppered separatory funnel should first be swirled to allow some mixing. Close the funnel and invert it with the stopper held in place and immediately open the stopcock. This should be done with the stopcock handle and funnel drain turned away from you, and with the other hand encompassing the barrel to keep the stopcock plug securely seated. (*Note*: Glass stopcocks should be lubricated: Teflon stopcocks should not.) Repeat until it is evident that there is no excessive pressure. Swirl again as the funnel is racked, immediately remove the stopper, and separate when appropriate. Always anticipate leakage by placing an adequate-sized beaker under the funnel.

When using large separatory funnels (>1 L) with volatile solvents, the vapor pressure can expel the stopper. It may be safer to work with a smaller funnel and perform several extractions.

Equipment Fabrication

1. Anneal all glassware prepared from glass-blowing (slow cooling to relieve thermal stresses).
2. Examine all glassware for stress under polarized light (crossed polarizers). Do not use glass that discloses stress patterns.
3. Examine glassware (especially used glass) for any combustible material before inserting it in a flame.
4. Silica dust and film from "blow out" can be a skin and respiratory system hazard; avoid contact.
5. During glass-blowing operations, wear eye protection for shielding from mechanical trauma and the bright sodium vapor glare.
6. Maintain constant vigilance for thermal burns.
7. Always fire-polish cut glass before use.

SELECTION OF GLASSWARE

Specific Requirements

There are many factors to consider when selecting glassware for laboratory needs. Some of these are:

1. Purpose of the glassware. Will it be used once and discarded, or will it be used heavily for years? Will it be subjected to heat, mechanical stress, or corrosive reagents?

2. Quality of glass required. In most cases, the type of glass to be used will be defined first by the way it is used and second by its cost. A glassware failure can be dangerous.
3. Cost.

Types of Glass

Glassware is not all composed of a homogeneous type of glass. There are different types of glass with different physical and chemical properties. They range from thin, inexpensive soda-lime glasses that can be thrown away after a single use, to expensive, pure silicon dioxide glasses for corrosive, high-temperature applications, with many types in between.

Glasses used in the laboratory are differentiated by their chemical composition and thermal treatment. In a school science laboratory, there are two types of glasses that are important: soda-lime glass and borosilicate glass. A typical soda-lime composition is 72% silica, 15% soda, 10% magnesia, 2% alumina, and 1% miscellaneous oxides. Borosilicate glass is formed by the addition of boron oxide (B_2O_3) to a soda-lime mixture. It is a "hard" glass that softens at a higher temperature than soda or "soft" glass. Borosilicate glass has excellent resistance to chemical, thermal, and mechanical shock. The common laboratory glasses, Pyrex and Kimax, are borosilicate glasses.

Because of its lower performance, the only soda glass used in the laboratory should be that in reagent bottles, stirring rods, and glass tubing.

Plastics

Plastic laboratory ware should be chosen on the basis of its chemical resistance. A table listing the chemical resistance of many plastics can be found in Chap. 10.

Corrosion Resistance

Silicate glasses are affected to some degree by most laboratory solutions. The following glass-corroding reagents are listed in order of decreasing reactivity: hydrofluoric acid, sodium hydroxide, sodium carbonate, ammonium hydroxide, phosphoric acid, and sulfuric acid. Hydrofluoric acid and hot phosphoric acid are so corrosive to glass that they should never be placed in glass containers.

Thickness

Glass apparatus comes in varying thicknesses for various purposes. When thicker glass is required, in vacuum and pressure applications or with large beakers, it should be used.

RECEIVING, STORAGE, AND DISPOSAL OF GLASSWARE

Like any fragile equipment, glassware should be carefully handled on receipt. Broken glass is frequently found in containers being unpacked. Instructors and students alike should assume that broken glass may be present when shipping cartons are being opened.

When received, each carton of glassware should be opened and the glass items within examined for cracked or nicked pieces. These pieces, if not replaced, may fail during use at some later date.

Special attention should be paid to the storage of glassware. Glassware should be stored in a well-lighted area on shelves designed for this purpose. Shelf edges should have a rim of sufficient height to prevent glass pieces from falling off the shelf. The following recommendations, if implemented, will provide a safe glass storage area:

1. Light pieces of glassware should be stored on upper shelves and heavy pieces on lower shelves.
2. Tall pieces of glassware should be stored at the back of the shelf and short pieces near the front.
3. All parties should be able to reach for glass pieces on the highest shelf without resorting to a stepladder or stool.
4. Glass rods and tubing should be stored in a horizontal position, and no piece should protrude over the edge of the shelf.
5. All aisles in the storage area should be kept clear of obstacles and debris at all times.

Broken glass should be disposed of in a specially marked container set aside for that purpose. It should not be disposed of in the normal trash receptacles. Janitors and cleaning personnel can be injured by broken glass in unmarked containers. When breakage occurs, the large glass pieces should be removed by sweeping them with a whisk broom into a dust pan. The small particles are then removed by wiping the entire area with wet cotton swabs or by vacuuming with a small vacuum cleaner.

4
ELECTRICAL HAZARDS

INTRODUCTION

During the past 25 years, the use of electrically powered apparatus in laboratories has increased more rapidly than that of any other kind of equipment. With the advent of microprocessor-controlled instrumentation, this trend is expected to continue. Electrical equipment is now used routinely for operations requiring heating, cooling, agitation or mixing, and pumping, as well as for a variety of instruments used in making physical measurements. The popularity of electrical apparatus is due to its convenience and relative safety. For instance, electrical heaters, rather than burners or other devices that have open flames, should be the only type of heat source present in operations involving the use of flammable materials. However, although the introduction of electrically powered equipment has resulted in a major improvement in laboratory safety, the use of this equipment does pose a new set of possible hazards for the unaware.

Laboratory workers should know the procedures for removing a person from contact with a live electrical conductor and the emergency first-aid procedures to treat someone who has suffered a serious electrical shock (see Chap. 13).

Remember, it is the current that hurts. The current running through the body at a point of contact depends on the voltage of the electrical source and the point of contact. An actual current of 25 mA can cause respiratory failure in a short time; 100 mA is quickly fatal. A moistened finger in contact with 110 V can allow a current of approximately 7–10 mA to flow through the body.

RECEPTACLES

All 110-V outlet receptacles in laboratories should be of the standard design that accepts a three-prong plug and provides a ground connection. The use of an old-style two-prong receptacle and an adapter that takes a three-prong plug is a less satisfactory alternative; however, no attempt should be made to bypass the ground. Old-style receptacles should be replaced as soon as feasible, and an additional ground wire added if necessary so that each receptacle is wired as shown in Fig. 4–1. If the use of an extension cord becomes necessary, standard three-conductor extension cords providing an independent ground connection should be used.

It is also possible to fit a receptacle with a ground-fault circuit breaker that will disconnect the current if continuity is lost in the system. Such ground-fault protection devices are frequently recommended for outdoor receptacles by local electrical codes and would be useful for selected laboratory receptacles where maintenance of a good ground connection is essential for safe operation.

Receptacles that provide electric power for operations in hoods should be located outside of the hood. This location prevents the production of electrical sparks inside the hood when a device is plugged in and also allows a laboratory worker to disconnect electrical devices from outside the hood. However, cords should not be allowed to dangle outside the hood in such a way that a worker could accidentally pull them.

Laboratory equipment that is to be plugged into a 110-V (or higher) receptacle should be fitted with a standard three-conductor line cord that provides an independent ground connection to the chassis of the apparatus (see Fig. 4–2). In some instances, "double-insulated" equipment with a two-conductor line cord may be adequate. All frayed or damaged line cords should be replaced before further use of the equipment is permitted; the annual inspection of all electrical cords is good practice. It is also desirable that equipment to be plugged into an electrical receptacle be fitted with a fuse or other overload-protection device that will disconnect the electrical circuit in the event the apparatus fails or is overloaded. This overload protection is particularly useful for apparatus

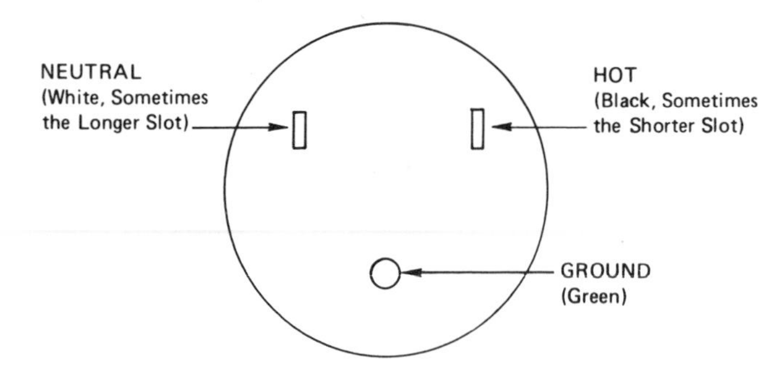

FIGURE 4-1. Standard wiring connection for 110-V receptacle (front view). (Reprinted from *Prudent Practices for Handling Hazardous Chemicals in Laboratory Safety Handbook,* with permission from the National Academy Press, Washington, D.C., 1981, p. 180.)

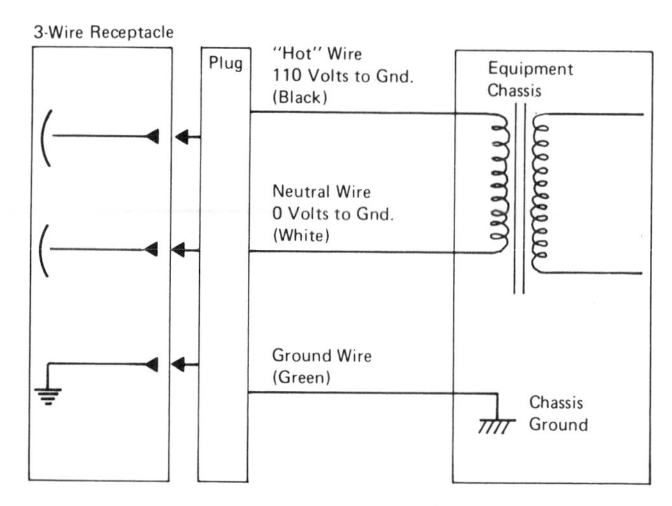

FIGURE 4-2. Standard wiring connection for line cord supplying 110-V electrical power to equipment. (Reprinted from *Prudent Practices for Handling Hazardous Chemicals in Laboratory Safety Handbook,* with permission from the National Academy Press, Washington, D.C., 1981, p. 181.)

likely to be left on and unattended for long periods of time [such as variable autotransformers (e.g., Variacs[RT] and Powerstats[RT]), vacuum pumps, drying ovens, stirring motors, and electronic instruments]. New or existing equipment not containing this overload protection can be modified to provide such protection.

EXPERIMENTAL WIRING

The hazards of experimental wiring lie principally in the temporary, possibly makeshift nature of components. Terminals are often exposed, and because work is being done on the project, no shielding or other protection is installed. Those doing experimental electrical work are usually quite competent, but there are a number of cases on record in which individuals have forgotten that a circuit was energized and touched a terminal, or made a principal contact. With certain types of electrophoresis equipment, accidental contact is not only possible, it has occurred in several cases.

GROUNDING

The grounding of laboratory electrical systems and electronic instrumentation is essential for the safe operation of a laboratory. All new laboratory installations are or should be equipped with grounded electrical outlets. Older laboratories still equipped with ungrounded outlets should be modernized as quickly as possible to provide an adequate and safe ground for electrical equipment. The upgrading of laboratory electrical systems deserves the highest priority and should be supported to the utmost by all instructors.

Before an instructor attempts to alter, modify, or otherwise work with a grounding system, he should seek the advice of qualified individuals. This is a task better left to school maintenance personnel.

Electrical systems and instrumentation can be checked for the adequacy of ground. This should be done by a qualified individual well versed in the testing of electrical systems.

MOTORS

Motor-driven electrical equipment used in a laboratory where volatile flammable materials may be present should be equipped with a nonsparking induction motor rather than a series-wound motor that uses carbon brushes. This applies to the motors used in vacuum pumps, mechanical shakers, and especially stirring motors, magnetic stirrers, and rotary evaporators. The speed of an induction motor operating under a load should not be controlled by using a variable autotransformer; such circumstances will cause the motor to overheat and might start a fire. There is no way to modify an apparatus that has a series-wound motor so that it will be spark-free. For this reason, many kitchen appliances (e.g., mixers and blenders) should not be used in laboratories where flammable materials may be present. Finally, it should be remembered that when other items of equipment (especially vacuum cleaners and portable electric drills) having series-wound motors are brought into a laboratory for special purposes, precautions (see Chap. 12) should be taken to ensure that no flammable vapors are present before such equipment in used.

VARIABLE AUTOTRANSFORMERS

A variable autotransformer (e.g., Variac[RT]) supplies some fraction of the total line voltage to a heater, light, or other electrical device in order to decrease its power output. New or existing variable autotransformers should be wired (or rewired) as illustrated in Fig. 4–3. If a variable autotransformer is not wired in this manner (older models were not), the switch on it may or may not disconnect both wires of the output from the 110-V line when in the off position. Also, if this wiring scheme has not been followed (especially the use of a grounded three-prong plug), even when the potential difference between the two output lines is only 10 V, each output line may be at a relatively high voltage (e.g., 110 and 100 V) with respect to an electrical ground. Because of these possibilities, whenever a variable autotransformer whose wiring is not definitely known to be acceptable is used, it is best to assume that either of the output lines could be at 110 V and capable of delivering a lethal electric shock.

The cases of all variable autotransformers have numerous openings to allow for ventilation, and some sparking may occur whenever the voltage adjustment knob is turned. Laboratory workers should be careful to locate these devices where water and other chemicals cannot be spilled on them (shock hazard) and where their movable contacts will not be exposed to flammable liquids or vapors (fire hazard). Specifically, variable autotransformers should be mounted on walls or vertical panels and outside of hoods; they should not just be placed on laboratory bench tops, especially those inside of hoods.

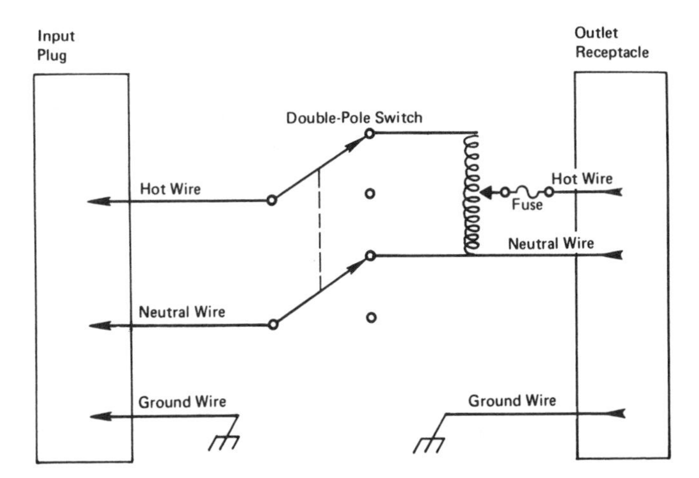

FIGURE 4–3. Schematic diagram of properly wired variable autotransformer. (Reprinted from *Prudent Practices for Handling Hazardous Chemicals in Laboratory Safety Handbook,* with permission from the National Academy Press, Washington, D.C., 1981, p. 188.)

WATER HAZARD

Whenever possible, electrical equipment should be located so as to minimize the possibility that water or chemicals could accidentally be spilled on it. If water or any chemical should accidentally be spilled on electrical equipment, the device should be unplugged immediately and not used again until it has been cleaned and inspected (preferably by a qualified technician). Water can also enter electrical equipment as condensation on equipment placed in a cold room or a large refrigerator. Cold rooms pose a particular hazard in this respect. The atmosphere in such rooms is frequently at a high relative humidity, and the potential for water condensate is significant. If electrical equipment must be placed in such areas, the condensation problem can be lessened (but not eliminated) by mounting the equipment on a wall or vertical panel. Other ways to minimize condensation are to use small resistive heaters or to pipe dry gas into the affected area. The potential for electrical shock in these rooms can be minimized by careful electrical grounding of the equipment and the use of suitable flooring material.

SERVICE

All laboratories should have access to a qualified technician who can make routine repairs to existing equipment and modifications to new or existing equipment so that it will meet reasonable standards for electrical safety.

With the exception of certain instrument adjustments that would be well described in an operator's manual, line cords of electrical equipment should always be unplugged before any adjustments, modifications, or repairs are undertaken. When it is necessary to handle a piece of electrical equipment that is plugged in, laboratory workers should first be certain that their hands are dry.

The laboratory worker may be exposed to a multitude of equipment hazards. A basic rule is not to attempt servicing equipment if you are inexperienced with electrical apparatus. Otherwise, read and follow the maintenance directions carefully, being sure that appropriate tools are available and safely insulated when needed.

Laboratory electrical supplies and other builtin systems are generally beyond the instructor's area of responsibility. The electrical system is commonly the responsibility of the building architect and design engineer. Electrical system maintenance is generally the responsibility of building maintenance personnel. However, every science teacher should learn the location of switches that would cut off power to his or her laboratory.

HIGH VOLTAGE

High voltages and charge storage make electronic equipment potentially lethal. Many types of electronic equipment operate at voltages in excess of 500 V. Photomultiplier circuits and laser power supplies can produce over 1000 V. Appropriate grounding of all electrical and electronic equipment is an absolute necessity. Properly designed instrumentation should have interlock switches that serve to cut off electricity to the unit if the cabinet is opened while the instrument is still on.

The potential remaining in condensers used in high-voltage work must be dissipated to ground in a safe way; otherwise, someone contacting the condenser or its electrodes may receive an extremely severe shock. Plastic shields can be placed over such equipment to prevent human contact, having switches interlocked with the shielding so that the equipment is deenergized before contact can take place. It is well known, however, that personnel frequently bypass safety devices such as the interlock switches in order to make adjustments. Sooner or later, a contact will be made.

EQUIPMENT FOR HAZARDOUS LOCATIONS

It is often necessary to operate electrical equipment in areas containing flammable vapors.

National Electrical Code: Article 500

Explosion-proof apparatus (switches, lights, motors, etc.) should be enclosed in a case capable of withstanding the interior explosion of a specified gas mixture and at the same time preventing ignition of the same mixture if it is surrounding the case. This capability is achieved by making the case of sufficient strength to withstand the explosion and also by designing all joints so that any flame will be quenched as gases pass through. Article 500 lists groups of gases or vapors, each requiring specific design characteristics. Groups for combustible dusts are also listed. Explosion-proof equipment designed for most hydrocarbons is not adequate for ethyl ether, hydrogen, or acetylene.

Intrinsically Safe Instruments

Another approach to safety with flammables has been developed principally in Europe. The term "intrinsically safe" means that the current level at any point is so low that a fault or short in the equipment could not generate a spark hot enough to ignite flammable vapors. In other words, the spark would be well below the minimum ignition energy required for ignition. This concept is growing, with respect to instrumentation. Of course, it would not be feasible for power equipment, nor is it applicable to atmospheres containing hydrogen, acetylene, etc.

Sealed Units

Sealed units such as mercury switches are inherently nonsparking, and so long as the wiring connecting them to the lines is unbroken and the insulation is good, they are relatively safe to use, particularly when only a slight possibility of hazardous vapor exists. However, they should not be considered acceptable for hazardous locations, and their use should be carefully evaluated.

Ventilated Enclosures

Explosion-proof motors are expensive, particularly in larger sizes, and the cooling air requirements for large explosion-proof motors are high. To meet this situation in large pump houses and other such locations, open motors are installed with ventilating ducts to take air from above the roof of the building. This air is forced or blown over or through the motor and then exhausted from the building. In this way, a high degree of safety is possible, with minimal cost for motors. Even with this arrangement, a possible hazard exists in that the ventilation system must be started before the motor can be started, and a suitable interlock is strongly recommended in order to make sure that this particular sequence prevails.

The same principle is applied for the purging of instrument cabinets in pilot plants and other areas where flammables are handled. A manometer should be installed to indicate that a positive differential is maintained.

OPERATING RULES

Some general rules for the safe operation of electrical systems and equipment in the lab can be summarized as follows:

1. One of the best ways to ensure safety is to use grounded equipment and outlets. Make sure that both the equipment and receptacle are grounded properly.

2. Use extension cords only when absolutely necessary and keep power cords as short as possible. Put them away after each use.
3. Whenever current leakage is detected in a piece of equipment (through a shock or tingling), pull the piece of equipment out of service immediately and send it out for repair. Insulation faults do not correct themselves; they grow worse with time.
4. Make sure that the addition of electrical equipment to existing electrical systems does not overload that system.
5. Only use equipment in good working condition and of adequate and proper design.
6. Inspect the equipment for frayed insulation and loose or broken wires.
7. Make sure that the bench area under the equipment is dry.
8. Make sure that the area around the equipment is free from flammables.
9. Be certain that all power switches are in the off position before plugging the appliance into an outlet. This prevents sparking at the plug.
10. Do not handle electrical equipment with wet hands or perspiring hands or while standing on a wet floor. Wear well-insulated shoes.
11. Unplug any electrical equipment before servicing it.
12. Do *not* jerk plugs from outlets.
13. Do *not* substitute a conductor for a fuse.
14. When checking electrical equipment for overheating, or if the equipment is suspected of having an electrical fault, check only with the back of one hand. If you receive an electrical shock, an involuntary muscular contraction will not cause you to grab the piece of equipment involved.
15. Do *not* store explosives or flammable liquids in refrigerators or near electrical motors.

Tables 4–1, 4–2, and 4–3 describe the effects of electric shock on the human body.

TABLE 4–1. Electrical Resistance of the Human Body

Mode of Contact	Resistance Value(Ω)
One dry finger on each electrode	100,000
One wet finger on each electrode	40,000
One salt/wet finger on each electrode	16,000
Tight grip on each electrode	1,200

Source: Reprinted with permission from *Modern Chemical Technology, Guidebook for Chemical Technicians,* R. L. Pecsok and K. Chapman (eds.), copyright American Chemical Society, Washington, D.C., 1970, pp. 4–3 and 4–4.

TABLE 4–2. Electrical Current Through the Human Body (For 110-V equipment, the application of Ohm's law, V=IR, yields the following results.)

Mode of Contact	R(Ω)	I(A)	I(mA)
Dry finger	100,000	0.0011	1.1
Wet finger	40,000	0.0028	2.8
Salty finger	16,000	0.0068	6.8
Tight grip	1,200	0.092	92.0

Source: Reprinted with permission from *Modern Chemical Technology, Guidebook for Chemical Technicians,* R.L. Pecsok and K. Chapman (eds.), copyright American Chemical Society, Washington, D.C., 1975, p. 72.

TABLE 4–3. Degree of Shock

Current (mA)		
ac (60 Hz)	dc	Effect
1	5	Threshold of sensation
1–3		Mild perception
6–9	70	Paralysis, inability to let go
25	80	Danger to life from heart/respiratory failure
100	100	Fibrillation, death
4000		Heart paralysis
>5000		Burning

Source: Reprinted with permission from *Modern Chemical Technology, Guidebook for Chemical Technicians,* R.L. Pecsok and K. Chapman (eds.), copyright American Chemical Society, Washington, D.C., 1975, p. 4–4.

SUGGESTED READINGS

Pecsok, R.L. and Chapman, K. (eds.), *Modern Chemical Technology, Guidebook for Chemical Technicians,* American Chemical Society, Washington, D.C., 1970.

American Chemical Society, *Safety in Academic Chemistry Laboratories,* Washington, D.C., 1985.

5

LABORATORY EQUIPMENT

INTRODUCTION

Most of the equipment used in laboratories is relatively safe in most situations. However, each type of device has particular hazards associated with its normal operation, and the possibility of equipment malfunction and failure must always be considered. Maintaining a safe laboratory requires knowledge of when a piece of equipment can be unsafe, what the limits of safe operation and failure modes are.

HEATING APPARATUS

Next to glass failure, the most common source of laboratory injury is the improper manipulation of heating apparatus, particularly gas burners. The use of steam-heated devices rather than gas or electric heating is generally preferred whenever temperatures of 100°C or less are required; these devices do not present shock or spark hazards and can be left unattended with the assurance that their temperature will never exceed 100°C.

Gas Burners

There are several types of gas burners. The tirril burner is capable of flame temperatures up to 900 K. The Fisher or Meker burner can produce temperatures up to 1000 K.

The barrels of gas burners often become very hot. Cooling time must be allowed before moving or manipulating the burner. Safety requires the use of wire gauze or asbestos-centered wire gauze between the burner flame and a glass flask or beaker, even the Pyrex type.

Any source of an open flame in a laboratory should not be used unless absolutely necessary. One of the few acceptable uses of a gas burner is glassware fabrication and modification.

Electrical Heating Equipment

Perhaps the most common types of electrical equipment in a laboratory are heaters. They are less hazardous than heating with an open flame, especially when heating flammable liquids. These electrical heaters include hot plates, heating mantles and tapes, oil baths, air baths, hot-tube furnaces, hot-air guns, soldering guns and irons, and hot-melt glue guns. Although they are inherently much safer than burners as laboratory heat sources, such devices can still pose both electrical and fire hazards if used improperly. Explosion-proof equipment is recommended.

General Rules for Using Electrical Heating Equipment

1. The actual heating element in any laboratory heating device should be enclosed in a glass, ceramic, or insulated metal case such that it is not possible for a laboratory worker (or some metallic conductor) to accidentally touch the wire carrying the electric current. This practice minimizes the hazards of electrical shock and of accidentally producing an electrical spark near a flammable liquid or vapor (see Chap. 12). This type of construction also diminishes the possibility that a flammable liquid or vapor will come into contact with the hot wire (whose temperature is frequently higher than the ignition temperature of many common solvents). If any heating device becomes so worn or damaged that its heating element is exposed, the device should either be discarded or the damage corrected before it is again used in the laboratory. Many household appliances (e.g., hot plates and space heaters) do not meet this criteria and, consequently, are not advisable for use in the laboratory.

2. The temperature of many laboratory heating devices (e.g., heating mantles, air baths, and oil baths) is controlled by the use of a variable autotransformer that supplies some fraction of the total line voltage to the heating element of the device. Older models may be wired in an unsafe manner that could cause significant shock and spark hazards. See Chap. 4 for a description of the correct wiring of variable autotransformers.

   The cases of all variable autotransformers have numerous openings to allow for ventilation, and some sparking may occur whenever the voltage adjustment knob is turned; laboratory workers should be careful to locate these devices where water and other chemicals cannot be spilled on them (shock hazard) and where their movable contacts will not be exposed to flammable liquids or vapors (fire hazard). Specifically, variable autotransformers should be mounted on walls or vertical panels and outside of hoods; they should not just be placed on laboratory bench tops, especially those inside of hoods.

3. Because the electrical input lines (even lines from variable transformers) to almost all laboratory heating devices may well be at 110 V with respect to any electrical ground, these lines should always be considered both as potential shock hazards (see Chap. 4) and as potential spark hazards (see Chap. 12). Thus, any connection from these lines to a heating device should be both mechanically and electrically secure and completely covered with some insulating material. Alligator clips should not be used to connect a line cord from a variable autotransformer to a heating device (especially an oil bath or air bath) because such connections pose a shock hazard and also may slip off, causing an electrical spark and, perhaps, contacting other metal parts to create a new hazard. All connections should be made by using either insulated binding posts or, preferably, a plug and receptacle combination.

4. Whenever an electrical heating device is to be left unattended for a significant period of time (e.g., overnight), it is advisable that it be equipped with a temperature-sensing device that will turn off the electric power if the temperature of the heating device exceeds some preset limit. Similar control devices are available that will turn off electric power if the flow of cooling water through a condenser is unexpectedly stopped. Such fail-safe devices, which can either be purchased or constructed by a qualified technician, prevent more serious problems (fires or explosions) that may arise if the temperature of an unattended reaction should increase significantly either because of a change in line voltage or because of the accidental loss of reaction solvent. These devices are also valuable accessories for use with stills employed to purify reaction solvents because such stills are often left unattended for significant periods of time.

Laboratory Hot Plates. Electrical hot plates are good for flat-bottomed vessels. Some hot plates are equipped with magnetic stirrers. Hot plates with exposed elements and those without continuously variable heat controls are dangerous and should not be used. Hot plates are available in sizes from 4 × 4 in. to several square feet of surface area. Laboratory hot plates are normally used when solutions are to be heated to 100°C or above and the inherently safer steam baths cannot be used as the source of heat.

As noted above, only hot plates with completely enclosed heating elements should be used in laboratories. Although almost all laboratory hot plates now being sold meet this criterion, many older ones pose an electrical-spark hazard arising from the on-off switch location on the hot plate, the bimetallic thermostat used to regulate the temperature of the hot plate, or both. Normally, these two spark sources are both located in the lower part of the hot plate in a region where any heavier-than-air (and mostly flammable) vapor escaping from a boiling liquid on the hot plate would tend to accumulate. In principle, these spark hazards could be alleviated by enclosing all mechanical contacts in a sealed container or by the use of solid-state circuitry for switching and temperature control, but in practice, such modifications would be difficult to incorporate in many of the hot plates now used in laboratories. Laboratory workers should be warned of the spark hazard associated with many existing hot plates (perhaps by attaching a warning label), and any new hot plates purchased should be constructed in a way that limits the possibility of electrical spark hazards. In addition to the spark hazard, old and corroded bimetallic thermostats in these devices can eventually fuse shut, thus delivering full current to the hot plate. This hazard can be avoided by wiring a fusible coupling into the line inside the hot plate so that if the device does overheat, the coupling will melt and interrupt the current.

Heating Mantles. Mantles are electrical elements wrapped in soft, insulating, heat-resisting cloth such as fiberglass. They are form-fitting for flasks and beakers. Mantles are most often used with round bottom flasks, reaction vessels, and distillation equipment. Mantle temperature is controlled by a variac.

As long as the fiberglass coating is not worn or broken or no water or other chemicals is spilled onto the mantle, these mantles pose no shock hazard to the worker. They are normally fitted with a male plug that fits into a female receptacle on an output line from a variable autotransformer to provide a connection that is mechanically and electrically secure.

Heating mantles should never be plugged directly into a 110-V line. Laboratory workers should be careful not to exceed the input voltage recommended by the manufacturer of a mantle because higher voltages will cause it to overheat, melting the fiberglass insulation and exposing the bare (and often red-hot) heating element. Note that the maximum recommended input voltage for a mantle being used with a dry flask is 10–20 V lower than that for a mantle being used with a flask containing a liquid. Some heating mantles are constructed by encasing a fiberglass mantle in an outer metal case that provides physical protection against damage to the fiberglass. If such metal-enclosed mantles are used, it is good practice to ground the outer metal case either by a three-conductor cord (containing a ground wire) from the variable autotransformer or by attaching one end of a heavy braided conductor to the mantle case and the other end to a good electrical ground such as a cold-water pipe. This practice provides the laboratory worker with protection against electric shock if the heating element inside the mantle is shorted against the metal case.

Oil Baths. Baths of water or oil can be used in conjunction with burners or electrical hot plates to further control temperature and heat distribution. However, many oils are flammable and some smoke when heated. Silicone oils are safer than natural organic oils because they are less flammable.

Electrically heated oil baths are often used as heating devices for small or irregularly shaped vessels or when a stable heat source that can be maintained at a constant temperature is desired. Three potential problems associated with oil baths are the flammability of the oil to be used, safety cutoff for excessive temperatures, and splash hazard from breakage of equipment during operation.

Selection of the kind of oil required is straightforward. For temperatures below 200° C, a saturated paraffin oil is often used; a silicone oil (which is more expensive) should be used for temperatures up to 300°C. Try to operate at least 20°C below the flashpoint of the oil.

An oil bath should always be monitored by a thermometer or other thermal-sensing device to ensure that its temperature does not exceed the flashpoint of the oil being used. For the same reason, oil baths left unattended should be fitted with explosion-proof (nonsparking) thermal-sensing devices that will turn off the electric power if the bath overheats. This type of electrical cutoff should also be used if the bath is to be operated within 20°C of the oil's flashpoint.

Bare wires should not be used as resistance devices to heat oil baths. They should be heated with a metal pan fitted with an enclosed heating element or with a heating element enclosed in a metal sheath (i.e., a knife heater, a tubular immersion heater such as a Calrod^{RT}, or its equivalent). The input connection for this heating element should be a male plug that will fit a female receptacle from a variable autotransformer output line.

Heated oil should be contained in either a metal pan or a heavy-walled porcelain dish; a glass dish or beaker can break and spill hot oil if accidentally struck with a hard object. The oil bath should be carefully mounted on a stable horizontal support such as a laboratory jack that can be easily raised or lowered without danger of the bath tipping over. A bath should never be supported on an iron ring because of the greater likelihood of accidentally tipping it over. Finally, oil baths heated above 100°C should have splash guards to prevent oil splatter caused by water (or some other volatile substance) falling into the hot bath. Such an accident can splatter hot oil over a wide area.

Air Baths. Electrically heated air baths are frequently used as a substitute for heating mantles when heating small or irregularly shaped vessels. Because of their inherently low heat capacity, such baths normally must be heated considerably above the desired temperature of the vessel being heated (typically 100°C or more). These baths should be constructed so that the heating element is completely enclosed and the connection to the air bath from the variable transformer is both mechanically and electrically secure. These baths can be constructed from metal, ceramic, or (less desirable) glass vessels. If glass vessels are used, they should be thoroughly wrapped with a heat-resistant tape so that if the vessel is accidentally broken, the glass will be retained and the bare heating element not exposed.

Heat Guns. Laboratory heat guns are constructed with a motor-driven fan that blows air over an electrically heated filament. They are frequently used to dry glassware or to heat the upper parts of a distillation apparatus during the distillation of high-boiling materials. The heating element in a heat gun typically becomes red hot during use and, necessarily, cannot be enclosed. Also, the on-off switches and fan motors are not usually spark-free. For these reasons, heat guns almost always pose a serious spark hazard. They should never be used near open containers of flammable liquids, in environments where appreciable concentrations of flammable vapors may be present, or in hoods that are being used to remove flammable vapors. Household hair dryers should not be used as substitutes for laboratory heat guns unless they have three-conductor or double-insulated line cords.

Heat Tapes. Heating tapes can be used to wrap tubing or odd-shaped vessels. Generally made of fiberglass or plastic-covered heating wires, tapes are used for relatively low-temperature applications. They, too, are controlled by variacs.

Incandescent Lamps. Electrical heating can also be supplied by incandescent light bulbs. Substances with low flashpoints, such as diethyl ether or petroleum ether, can be safely heated with incandescent bulbs mounted in cylindrical supports. Again, they can be controlled with variacs. Carbon disulfide is extremely dangerous and can easily be heated to its flashpoint by a 40-W incandescent lamp.

Heating Techniques

To be safe, the boiling of liquids must be smooth. Liquids often superheat in spots and then vaporize, resulting in "bumping." This may cause liquid to overflow or shoot out through a condenser tube. Bumping can be avoided by adding sharp-edged inert pieces of glass, unglazed porcelain, or clay, called boiling chips. These chips provide surfaces on which the liquid can produce fine bubbles of gas. However, boiling chips should never be added to hot liquids as a sudden outburst of vapor may drive the liquid out of the container. Cool the liquid before adding boiling chips.

Boiling chips will not work when boiling liquids under vacuum. To prevent bumping, insert a thin capillary tube (with one end exposed to the air or connected to an inert gas line) in the bottom of the flask. Under the vacuum, a stream of bubbles will be drawn from the tube into the liquid, thereby facilitating the boiling action. The stream of gas can be controlled by a screw clamp and short length of rubber tubing. See Fig. 5–1 on how to boil liquid under a vacuum.

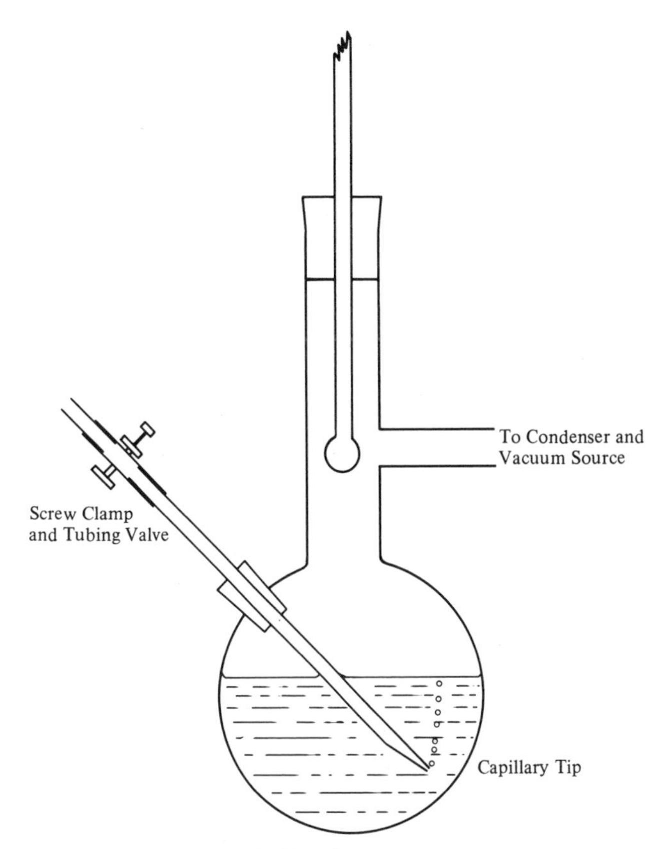

FIGURE 5–1. Boiling liquid under a vacuum.

All heating operations should be approached with extreme caution. Glass containers under a vacuum should be shielded with a plastic shield or wire screen. Many liquids used in laboratories are flammable and/or toxic. Care must be taken particularly when they are heated since any vapors will also be toxic or flammable. When in doubt as to the appropriate technique for heating a sample, consult an immediate supervisor.

In case of fire, first shut off the gas or pull the electrical connection. Then, if possible, suffocate the fire with a nonflammable object such as a beaker, watchglass, or evaporating dish. If the fire cannot be easily snuffed, use an appropriate extinguisher. Be careful not to knock over the container or spill the contents, thus spreading the fire.

STIRRING AND MIXING DEVICES

The stirring and mixing devices commonly found in laboratories include stirring motors, magnetic stirrers, shakers, small pumps for fluids, and rotary evaporators for solvent removal. These devices are typically used in laboratory operations performed in a hood, and it is important that they be operated in a way that precludes the production of electrical sparks within the hood (see Chap. 9). Furthermore, it is important that, in the event of an emergency situation, a laboratory worker be able to turn such devices on or off from outside the hood. Finally, heating baths associated with these devices (e.g., baths for rotary evaporators) should also be spark-free and capable of control from outside the hood.

Only spark-free induction motors should be used to run stirring and mixing devices; any motor that may produce sparks during its startup or running cycle should not be used. Fortunately, the motors in most of the currently marketed stirring and mixing devices meet this criterion. However, the on-off switches and rheostat-type speed controls of many of them do not meet this criterion because the switch or rheostat has exposed contacts that can produce an electrical spark any time a change in controls is made. This problem is particularly true of many of the magnetic stirrers and rotary evaporators currently being sold. One effective solution is the modification of the stirring or mixing device by removing any switches located on the device and inserting a switch in the line cord near the plug end. As the electrical receptacle for the plug should be outside the hood, this modification will ensure that the switch will also be outside of the hood. The speed of an induction motor operating under a load should not be controlled by a variable autotransformer (see Chap. 4).

Because stirring and mixing devices (especially stirring motors and magnetic stirrers) are often operated for fairly long periods without continual attention (e.g., reaction mixtures that are stirred overnight), the consequences of stirrer failure, electrical overload, or blockage of the motion of the stirring impeller should be considered. Many laboratory explosions have been caused by the failure of a stirring device during a reaction.

It is good practice to attach a stirring impeller to the shaft of the stirring motor by using lightweight rubber tubing. Then, if the motion of the impeller becomes blocked (e.g., by the formation of a copious precipitate), the rubber simply twists until it breaks, rather than either the motor stalling or breaking the glass apparatus into which the stirring impeller extends. It is also very desirable (but unfortunately not very common) that stirring motors to be left unattended be fitted with a suitable fuse or thermal-protection device.

ELECTRONIC INSTRUMENTS

Most modern electronic instruments are fitted with a line cord containing a separate ground wire for the chassis and with a suitable fuse or other overload protection. Any existing instrument lacking these features should be modified to incorporate them. As with any other electrical equipment, special precautions should be taken to avoid the possibility that water or other chemicals could be spilled on these instruments.

Under most circumstances, any repairs to, adjustments of, or alterations in such instruments should be made by a qualified technician. If laboratory workers do undertake repairs, the instrument line cord should always be unplugged before any disassembly begins. Certain instrument adjustments can be made only when the instrument involved is connected to power. Laboratory workers should not undertake such adjustments without supervision unless they have received specific prior instruction. This precaution is particularly important with instruments that incorporate high-voltage circuitry (such as oscilloscopes, spectrometers with photomultiplier tubes, and most equipment that uses vacuum-tube circuitry).

See Chap. 4 for a summary of laboratory electrical hazards.

CENTRIFUGES

Manufacturer's instructions must be followed at all times, and centrifuges should be serviced regularly. High-speed centrifuges should have their use logged and components derated or replaced according to the manufacturer's recommendations. It is important to prevent bucket corrosion or detect it as early as possible. Centrifuges and their accessories must be examined regularly for signs of corrosion, cracks, or undue wear. Periodically, the rubber cushions and buckets that hold them should be removed, disinfected, and finally rinsed in tap water and stored in inverted form. The inside of the bowl should be swabbed with a suitable disinfectant at appropriate intervals.

The safest containers are screw-capped, heavy-duty glass bottles or polypropylene containers that are autoclavable. The container should never be filled to more than two-thirds of its volume, but should be effectively capped. When you are transferring a specimen to the container, the tip of the pipet should be well inserted and not touch the rim.

Operation

Before use, care must be taken to ensure that loaded buckets and trunnions are properly balanced. Balancing liquid must not be added directly to centrifuge buckets, but added in a container similar to those containing the material to be centrifuged. The rubber cushions must be correctly seated in the buckets, trunnions properly in place, and multiple containers loaded symmetrically. After use, the lid must not be opened until the rotating head has come to rest. Rotation must never be braked by hand. The use of centrifuges with interlocked covers is recommended, and centrifuges without such protective interlocking devices should be phased out.

Category B pathogens (see Chap. 3) should be centrifuged in sealed buckets or totally wrapped in a sealed bag. They should only be handled in an exhaust-protective-type cabinet.

Breakages in the Centrifuge

If breakages occur when centrifuging bacterial cultures, the centrifuge must be stopped and at least 30 min must pass for the aerosols to settle. Before opening the centrifuge, slip on strong, disposable gloves. The contents of the centrifuge, including whenever possible the buckets, rotor, and trunnions, should be transferred to a metal bucket or autoclave bag ready for autoclaving. The head and inside of the bowl must be thoroughly swabbed with a noncorrosive disinfectant such as a phenolic or 4% formaldehyde solution (10% formalin) and left for 1 h before they are swabbed out with water and dried. All swabs must be treated as infected waste and sterilized before disposal.

AUTOCLAVES

Autoclaves should only be operated by authorized persons who have received full instructions from a senior staff member. All autoclaves must have passed a hydraulic test for which a test certificate has been issued. The pressure relief valve must be set at a pressure not exceeding the design pressure of the equipment. This should be locked to prevent unauthorized operation, but periodically it must be checked to ensure that it operates satisfactorily. Gloves should be worn for this purpose. Gas and electrically heated autoclaves must be filled to the correct level with water before loading, and the door or lid must be inspected to see if there has been any visible damage to the seal. On loading, care must be taken to ensure that vents or safety valve openings are not obstructed.

When autoclaving screw-capped glass vessels, the caps must first be loosened to avoid explosions when heating or cooling down. Stoppered bottles should have their tops removed or prevented from closure during autoclaving by an insert, such as a piece of string in the neck of the bottle. Operating instructions should be posted in the immediate vicinity of each autoclave and followed at all times.

When sterilization is complete, the source of heat must be turned off and the autoclave allowed to cool to below 100°C (at which point, the pressure gauge should indicate zero pressure) before the exhaust valve may be opened. When liquid media are involved, the temperature in the chamber must fall to 80°C before any attempt is made to remove the items. The sterilization of single large volumes of media should be avoided because these require a long time interval to cool satisfactorily.

A protective visor should always be worn when opening an autoclave, and the equipment should be approached from the hinge side in order to avoid scalding by the release of residual steam. At the same time, the door should be kept between the operator and the contents of the chamber whenever possible as an additional precaution against the explosion of individual vessels.

HOMOGENIZERS

The homogenization of contaminated material may result in the production of bacterial aerosol. To minimize this hazard, the homogenizing equipment should be contained in a glove box or strong bag whenever appropriate. The homogenization of material containing very high numbers of pathogens is to be avoided and when unavoidable should take place in a fume hood or glove box.

REFRIGERATORS

The potential hazards posed by laboratory refrigerators are in many ways similar to those of laboratory drying ovens. Because a satisfactory arrangement for continuously venting the interior atmosphere of a refrigerator almost never exists, any vapors escaping from vessels placed in one will accumulate. Thus, the atmosphere in a refrigerator could contain an explosive mixture of air and the vapor of a flammable substance or a dangerously high concentration of the vapor of a toxic substance or both. (The problems of toxicity are aggravated by laboratory workers who place their faces inside a refrigerator while searching for a particular sample, thus ensuring the inhalation of some of the atmosphere from the refrigerator interior.) As noted in Chap. 6, laboratory refrigerators should never be used for the storage of food or beverages. There should be no potential sources of electrical sparks on the inside of a laboratory refrigerator. In general, it is preferable when purchasing a refrigerator for a laboratory to select a "flammable storage" model that has been designed by the manufacturer for that specific purpose. If this is not possible, all new or existing refrigerators should be modified to remove all spark sources by (1) removing any interior light activated by a switch mounted on the door frame and the switch, (2) moving the contact of the thermostat controlling the temperature to a position outside the refrigerated compartment, and (3) moving the contacts for any thermostat present to control fans within the refrigerated compartment to the outside of the refrigerated compartment. Although a prominent sign warning against the storage of flammable substances could be permanently attached to the door of an unmodified refrigerator, this alternative is less desirable than the modification of the equipment by removal of all spark sources from the refrigerated compartment. "Frost-free" refrigerators are not advisable for laboratory use because of the many problems associated with attempts to modify them. Many of these refrigerators have a drain tube or hole that carries water (and any flammable material present) to an area adjacent to the compressor and, thus, they present a spark hazard; the electric heaters used to defrost the freezing coils are also a potential spark hazard. Laboratory refrigerators should be placed against fire-resistant walls, have heavy-duty cords, and preferably be protected by their own circuit breaker.

Uncapped containers of chemicals should never be placed in a refrigerator. Containers of chemicals should be capped so as to achieve a seal that is both vapor-tight and unlikely to permit a spill if the containers are tipped over. Caps constructed from aluminum foil, corks, corks wrapped with aluminum foil, or glass stoppers often do not meet all of these criteria; the use of such methods for capping containers should be discouraged. The most satisfactory temporary seals are normally achieved by using containers that have screw-caps lined with either a conical polyethylene or Teflon[RT] insert. The best containers for samples to be stored for longer periods of time are sealed, nitrogen-filled glass ampules.

The placement of potentially explosive or highly toxic substances (see *Academic Laboratory Chemical Hazards Guidebook,* Chap. 1) in a laboratory refrigerator is strongly discouraged. If this precaution cannot be taken, then a clear, prominent warning sign should be placed on the outside of the refrigerator door. The length of refrigerator storage of such material should be kept to a minimum.

VACUUM PUMPS

Distillations or concentration operations involving significant quantities of volatile substances should normally be performed by using a water aspirator or steam aspirator, rather than a mechanical vacuum pump. However, the distillation of less volatile substances, removal of final traces of solvents, and some other operations require pressures lower than those obtained by using a water aspirator and are normally performed with a mechanical vacuum pump. The output line from the system to the vacuum pump should be fitted with a cold trap to collect volatile substances from the system and minimize the amount that enters the vacuum pump and dissolves in the pump oil; the use of liquid nitrogen or liquid air in such traps, however, can lead to a flammability hazard (see Chap. 8). The possibility that mercury will be swept into the pump as a result of a sudden loss of vacuum can be minimized by placing a Kjeldahl trap in the line at the pump.

The output of each pump should be vented to an air exhaust system (see Chap. 9). This procedure is essential when the pump is being used to evacuate a system containing a volatile toxic or corrosive substance (failure to observe this precaution would result in pumping the substance into the laboratory atmosphere); it may also be necessary to scrub or absorb vapors. Even with these precautions, however, volatile toxic or corrosive substances may accumulate in the pump oil and, thus, be discharged into the laboratory atmosphere during future pump use. This hazard should be avoided by draining and replacing the pump oil when it becomes contaminated. The contaminated pump oil should be disposed of by following standard procedures for the safe disposal of toxic or corrosive substances (see *Academic Laboratory Chemical Hazards Guidebook,* Chap. 5).

Belt-driven mechanical pumps with exposed belts should have protective guards. Such guards are particularly important for pumps installed on portable carts or the tops of benches where laboratory workers might accidentally entangle clothing or fingers in the moving belt, but they are not necessary for pumps located at a height of at least 7 ft above the working surface or in an enclosed cabinet.

DRYING OVENS

Electrically heated ovens are commonly used in the laboratory to remove water or other solvents from chemical samples and to dry laboratory glassware before its use. With the exception of vacuum drying ovens, these ovens rarely have any provision for preventing the discharge of the substances volatilized in them into the laboratory atmosphere. Thus, it should be assumed that these substances will escape into the laboratory atmosphere and could also be present in concentrations sufficient to form explosive mixtures with the air inside the oven (see Chap. 12).

Ovens should not be used to dry any chemical sample that has even moderate volatility and might pose a hazard because of acute or chronic toxicity unless special precautions have been taken to ensure continuous venting of the atmosphere inside the oven. Thus, most organic compounds should not be dried in a conventional laboratory oven.

Glassware rinsed with an organic solvent should not be dried in an oven. If such rinsing is necessary, the item should be rinsed again with distilled water before being placed in an oven.

Because of the possible formation of explosive mixtures by volatile substances and the air inside an oven, laboratory ovens should be constructed so that their heating elements (which may become red hot) and temperature controls (which may produce sparks) are physically separate from their interior atmospheres. Many small household ovens and similar heating devices do not meet these requirements and, consequently, should not be used in laboratories. Existing ovens not meeting these requirements should either be modified or have a sign attached to their oven doors warning workers that flammable materials should not be placed within. Some safety groups suggest that every laboratory oven be modified by placing a blow-out panel in its rear wall so that any explosion within the oven will not blow the oven door and oven contents into the laboratory. NFPA standards call for blow-out vents on ovens handling flammable substances.

Thermometers containing mercury should not be mounted through holes in the tops of ovens so that the bulb of mercury hangs into the oven. Bimetallic strip thermometers are a preferable alternative for monitoring oven temperatures. Should a mercury thermometer be broken in an oven of any type, the oven should be turned off and all mercury removed from the cold oven (see Chap. 12).

DISTILLATION EQUIPMENT

Superheating and sudden boiling frequently occur when distilling under vacuum. Therefore, it is important that the assembly be secure and the heat be distributed more evenly than is possible with a flame. This can best be done with a mantle heater, ceramic cavity heater, or nonflammable liquid bath. Silicon oil or another suitable high-boiling oil can be used if heated on a hot plate. Hot water or steam should be used when practical. Evacuate the assembly gradually to minimize the possibility of "bumping." A standing shield should be in place for protection in the event of an implosion. An additional thermometer should be inserted very near the center bottom of the distilling flask to warn of a dangerous, exothermic decomposition. Avoid heating above the recommended temperature. Boiling chips are of little value in a vacuum distillation. After finishing a vacuum distillation, cool the system before slowly bleeding the air, because air may induce an explosion in a hot system.

Ethers must never be distilled unless known to be free of peroxides. Most ethers, including cyclic ethers, form dangerously explosive peroxides on exposure to air and light. The peroxides also may change the course of the planned reaction. Unsaturated hydrocarbons and other reagents can also form peroxides (see *Academic Laboratory Chemical Hazards Guidebook,* Chap. 1).

When conducting steam distillation, take care not to run the steam in at too great a rate for the condenser. Overfilling the flask should be avoided by trapping condensate in the entering steam line and by moderately heating the flask to prevent excessive condensation.

6
LABORATORY RULES

INTRODUCTION

An element of danger is always present in chemical, biochemical, and physical science laboratories. The hazard is not as serious in places where only routine procedures are carried out as in a research laboratory where the scale of work often involves quite large setups and where new chemicals, reactions, and equipment with corresponding new hazards are usual features. *Everything* in a laboratory, unless you know it to be perfectly harmless, should be treated with respect. Some consideration, from the safety aspect, should be given to every experiment before it is begun. Try to decide what may go wrong and then consider what emergency measures are to be taken if something does.

When goggles, screens, and protective clothing such as gloves are provided, they should be used, but the mere fact of using them does not mean that undue risks can be taken. Fire is probably the major laboratory hazard because of the rapidity with which it can spread with flammable solvents around. Remember that many fires are caused by smoking that should be absolutely forbidden near flammable solvents or vapors. Due care and observation of precautions will always be the main line of accident prevention and personal protection.

The following are safety rules that will greatly assist in reducing the more common accidents in the laboratory. Most are common-sense procedures and are no more than part of a generally thoughtful approach to laboratory safety.

RULES FOR STUDENTS

1. Never work alone in the laboratory chemical storage area.
2. Do not fool around in the laboratory. Horseplay and pranks can be dangerous.
3. Do not eat, drink, smoke, or chew gum or tobacco in laboratories or chemical storage areas. Do not keep food or beverages in the laboratory, including refrigerators. Do not use laboratory glassware as food or beverage containers. Potassium cyanide looks just like sugar.
4. Observe good housekeeping principles in the laboratory.
5. Never pipet by mouth.
6. Wash hands before and after work in the laboratory, and after spill cleanups.
7. Never leave heat sources unattended (e.g., gas burners, hot plates, heating mantles, sand baths, etc.).
8. Never lean into the fume hood.
9. Analyze new lab procedures in advance to pinpoint hazardous areas.
10. Always inform co-workers of plans to carry out hazardous work.
11. Do not perform unauthorized experiments.
12. Read all procedures. Look for possible hazards. Work defensively. It is insufficient to follow safety procedures by habit; always consider the implications of what you do.

RULES FOR INSTRUCTORS AND SUPERVISORS

1. Be an example:
   (a) Observe all rules.
   (b) Wear protective clothing and equipment when required.
   (c) Promote safety.
2. Look for unsafe conditions.
3. Make inspections often.
4. Correct safety problems immediately.
5. Maintain discipline and enforce rules.
6. Assume responsibility for visitors and be sure that they follow the laboratory safety rules.

7. Review all laboratory experiments for possible health, safety, and environmental problems before the experiments are assigned to students.
8. Maintain a library of publications on laboratory safety and make it readily available to anyone in the laboratory, thereby encouraging its use. MSDS's for all chemicals should be kept in an organized file.
9. Permit equipment to be operated only by those who have been appropriately trained in its proper use.
10. Inform lab occupants about the alarm bell and what to do if it sounds.
11. Supervisory personnel should think "safety." Their attitude toward fire and safety standard practices is reflected in the behavior of their entire staff.
12. Analyze accidents to prevent repeat performances.
13. A safety program is only as strong as the worker's will to take the correct actions at the right time.

Planning

1. Record who worked with what, when, and for how long in order to allow meaningful retrospective contamination studies.
2. Conduct regular in-house safety and health inspections, with an emphasis on improvement rather than guilt.
3. Carry out regular fire or emergency drills with critical reviews of the results.
4. Have actions preplanned in case of an emergency (e.g., what devices should be turned off, which escape route to use, a personnel meeting place outside the building, a person designated to authorize reentry into the building).
5. Lab personnel should have recent training in first aid, CPR, etc.

PROPER CLOTHING

1. Always wear appropriate clothing and laboratory protective gear.
2. Never wear sandals or shorts. Exposure of legs and feet to spilled chemicals is a main cause of chemical burns, second only to burns on the hands.
3. Avoid wearing loose-fitting clothes (e.g., sleeves, full-cut blouses, neckties, dangling jewelry, etc.).
4. Confine long hair in the laboratory.

SAFETY WEAR AND PROTECTIVE EQUIPMENT

1. ANSI-approved (or equivalent standard) eye or face protection should be worn continuously. Safety glass resists shattering when struck by objects. Contact lenses should not be worn. Use goggles when contact lenses must be worn for medical reasons.
2. Gloves should be worn that will resist penetration by the chemical being handled and that have been checked for pinholes, tears, or rips.
3. Wear a laboratory coat or apron to protect skin and clothing from chemicals.
4. Footwear should cover feet completely; no open-toe shoes should be worn.
5. Use goggles, shields, gloves, etc. whenever indicated. A wide variety of personal protection equipment is available.
6. Know the location and proper use of safety equipment: showers, eyewash, fire blankets, fire extinguishers, etc.
7. Use the laboratory safety equipment provided.
8. Protection should be provided for not only the lab worker, but also a lab partner working nearby.

FACILITIES AND EQUIPMENT

1. Keep separate containers for trash and broken glass.
2. Never block any escape routes and plan alternate ones.
3. Never block a fire door in the open position.
4. Never store materials in lab or storage area aisles.
5. Place fire extinguishers near an escape route, not in a "dead end."
6. Regularly maintain fire extinguishers, and records, and train personnel in the proper use of extinguishers in actual fire situations.
7. Acquaint personnel with the meaning of "Class A fire," "Class B fire," etc., and how they relate to fire extinguisher use.
8. Instruct lab personnel in the proper use of the eyewash fountain, emphasizing rolling of the eyeballs and turning eyelids "insideout."

9. Ensure that eyewash fountains will supply at least 15 min of water flow.
10. Regularly inspect safety showers and eyewash fountains for proper operation and keep records of inspections. Safety showers should be inspected daily by the room occupant to make sure the supply valve is on, the chain is not tangled, tied up, or otherwise inoperative, and the chain guide is not used as a coat rack.
11. Keep up-to-date emergency phone numbers posted next to the telephone.
12. Regularly check the hood for proper draft; also check that exhaust air from an external hood vent is not redrawn into the room's air.
13. Sample breathing air space for the measurement of possible contaminants and keep good records.
14. Secure all compressed gas cylinders when in use and transport them on a hand truck.
15. Install chemical storage shelves with lips and never use stacked boxes in place of shelves.
16. Only use explosion-proof refrigerators for lab storage.
17. Have appropriate equipment and materials available for spill control; replace them when they become dated.
18. Regularly inspect fire blankets for rips and holes and keep good records of such inspections.
19. All moving belts and pulleys should have safety guards.

Glassware

1. Lubricate tubing before inserting it into stoppers. Tubing may break and pierce the hand.
2. Avoid using chipped or cracked glassware.
3. Vacuum dessicators fabricated from glass should always be protected with a wire mesh guard or a similar device to contain fragmented glass in the event of an implosion, and also to protect them from being accidentally hit by a hard object.
4. In vacuum distillations or when evaporating solvents in a rotating vacuum evaporator, only *round-bottomed* flasks free from flaws ("stars") should be used. Nonspherical flasks are subject to uneven stresses when under a vacuum and are likely to implode.
5. Tape all Dewar flasks.

FIRE PREVENTION

One fire, or one accident, is one too many!

1. Know how to use the various types of fire extinguishers, the gas mask, and the fire alarm. Take time to read instruction labels before an emergency arises. The fire department will be glad to demonstrate any of this equipment on request.
2. Firefighting and emergency equipment, such as hose cabinets, electrical panel switch boxes, extinguishers, blankets, and laboratory module escape doors, should not be blocked.
3. Stairwell doors should be kept shut at all times. A closed stairwell is vitally important in safeguarding personnel and buildings: It serves as an avenue of escape; it minimizes the possibility of hot gases rushing upward to ignite an entire structure; it prevents the rapid dispersion of toxic fumes from broken or leaking containers; it provides for the safe entrance of emergency personnel.
4. Before leaving the lab, turn off fans, lights, and all unused laboratory equipment. If you want experimental equipment to operate overnight, you must placard it with a DO NOT DISTURB sign and write in your name and home telephone number. Start such an operation soon enough before leaving the building in the evening to assure that the equipment involved is functioning properly and that surfaces, wiring, etc., are not becoming unusually hot.
5. Do not smoke where NO SMOKING signs are placed. If you are not familiar with a particular laboratory room, you should always ask permission to smoke. Vapors of flammable solvents, such as ether and benzene, float along the floor for as much as 35 ft, and therefore, it is quite important to observe all NO SMOKING signs whether a flammable solvent appears to be present or not.
6. Smocks, coats, trousers, or undergarments made of synthetic fabrics not blended with cotton are known to accumulate static electricity in dry weather (particularly in winter) and will discharge with a spark near metallic objects. They should not be worn when working with highly flammable solvents with low flashpoints. The use of commercially available laundry softeners on these fabrics minimizes their tendency to collect static charges.

7. Flammable solvents should not be kept in refrigerators unless the refrigerators have been explosion-proofed.
8. Flammable solvents should never be stored on reagent shelves higher than eye-level and only in small (less than 500 ml) bottles readily grasped in one hand. Solvents stored at higher levels in larger bottles can be accidentally dropped and ignited, causing a major catastrophe since fire will spread quickly and extensively given the large volumes of splashing liquid.
9. Low boiling solvents such as carbon disulfide, ether, and petroleum ether should not be filled to the top of a closed container. A 30% expansion space must be allowed to absorb the pressure due to expansion of the vapors with temperature fluctuations.
10. Transfer flammable solvents only by pouring through a stainless steel funnel, to which ground leads have been attached.
11. A hot plate should never be used for heating flammable solvents because the ignition temperatures of most solvents are below the maximum temperature of hot plate surfaces. Some hot plates are equipped with a thermostatic control that adds to the hazard by producing a make-and-break spark.
12. Excessive overstocking of hazardous chemicals (flammable liquids, lachrymators, cylinders of toxic or flammable gases, explosive water-reactive materials, etc.) is discouraged. Although it may be more inconvenient and expensive to order small quantities more frequently, it does minimize fire risk to the building and life hazard to emergency response team members.
13. Empty chemical containers should be thrown out into trash baskets with their caps or stoppers off and only after being thoroughly rinsed with water. Explosions have been known to occur when capped bottles and cans with residual solvents were incinerated.
14. Do not extend the receptacle circuit in your laboratory by the use of multiple cube taps or extension strips. If your receptacle circuit system is inadequate, request that the maintenance department correct this deficiency.
15. Powerstat and centrifuges should be turned off with the ON-OFF *switch* when not in use. These apparatus are wired so that it is possible for the full-line voltage to exist across the terminals when only the rheostat dial is in the OFF position, and they may be accidentally energized.
16. New electrical codes require all new electrical laboratory apparatus to have ground wire leads. Connect these to three-prong *grounded* receptacles directly or use the adapter provided. Be sure the adapter ground lead is actually connected to the screw holding the front plate of the receptacle.
17. Learn the location of power circuit switches. With preknowledge of the switch locations to your own receptacle circuits or the "walk-in" box receptacles, you may shut down electrical power from a position remote from the immediate hazard area.
18. Cold rooms and other walk-in boxes can be potential bombs because they are *not* explosion-proof. Procedures requiring flammable or toxic substances should not be performed within such a room unless no possible alternative exists. If this kind of method must be used, post suitable warning signs on the door and make certain that all unnecessary electrical equipment is shut down. Electrical apparatus should be shut off at the switch before disengaging the plug. Know the location of the corridor panel box switch controlling the room's power receptacle system so that in an emergency, the electricity can be turned off from a remote location.
19. The lab vacuum system should be kept free of contamination. Solvents can break down the compressor oil seal and also create a potential explosion hazard. A dry-ice trap should be used to condense water-insoluble vapors, calcium chloride to remove moisture, and sodium hydroxide to neutralize acid vapor. *By using a water aspirator, these hazards are obviated.*
20. Sink and floor drain traps should be kept filled with water to prevent toxic or explosive concentrations of gas vapors from entering the room. This can be a problem with floor drains; because they are seldom used, all the water tends to evaporate from the trap.

PURCHASE, USE, AND DISPOSAL OF CHEMICALS

These rules will be helpful, but it is important that the basic reasons why certain chemicals are explosive or poisonous be well understood by all those who come into contact with or handle these materials.

Safe Storage of Chemicals

1. If possible, purchase chemicals in quantities small enough to be used in one class lesson.
2. Label all chemicals accurately with the date of receipt, or preparation, initialed by the person responsible, and any pertinent precautionary information on handling.
3. Generally, bottles of chemicals should not remain unused on shelves in the lab for more than one week, unused in the storeroom near the lab for more than one month, or unused in the main stockroom for more than one year.
4. Properly store flammable liquids in small quantities.
5. Do not use fume hoods as chemical storage areas.
6. Do not allow reagent areas to become cluttered.
7. Do not store reagents and/or apparatus on lab benches and keep lab shelves organized.

Safe Use and Handling of Chemicals

1. Chemicals in any form can be safely stored, handled, and used if their hazardous physical and chemical properties are fully understood and the necessary precautions, including the use of proper safeguards and personal protective equipment, are observed.
2. Never open a reagent package until the label has been read and completely understood. Know the hazards of the materials you are using.
3. Obtain and read the MSDS for each chemical involved in an experiment before beginning it.
4. Unless otherwise properly instructed, *never* attempt to smell or taste a chemical.
5. Avoid skin contact with chemicals.
6. Chemicals accidentally contacting hands should be rinsed away with cold water before a hot water soap cleansing. Hot water used first is more likely to open the pores, thus providing ready entrance through your skin.
7. If you have been burned with an acid or alkali, flush the area with a lot of water. If the burn is serious, call for emergency medical aid and continue water treatment until a physician arrives.
8. When an unusually toxic chemical requiring a special antidote and prompt first aid is being handled, the experimenter should be able to advise a physician of the situation's particulars. Examples of such chemicals are hydrogen cyanide, snake venom, nitrogen mustard, hydrofluoric acid, etc.
9. Never place reactive chemicals (in bottles, beakers/flasks, wash bottles, etc.) near the edges of a lab bench.
10. Do not mix chemicals in a sink drain.
11. Use a fume hood when working with volatile substances.
12. Do not remain in an area where unusually dominant odors of chemicals exist unless you are assured that the gas or vapor is not hazardous. Do not linger where chemicals have been spilled. Close the door, put on a gas mask, and attempt to control and/or clean up the chemicals. Be aware that many of the toxic gases and solvent vapors are rather pleasant, for example, carbon tetrachloride, phosgene (a chemical warfare gas which smells like new mown hay) and hydrogen cyanide has the odor of bitter almonds.
13. Electric power failure or shutdown puts exhaust hoods out of business. Cylinders of toxic or flammable gas must be turned off; reactions producing toxic fumes must be curtailed or stopped; bacteriological or virological techniques producing pathogenic aerosols must be halted. A leaking cylinder is always a threat: Stop the leak or dispose of the cylinder.
14. Purified preparations of proteolytic enzymes that may appear innocuous are hazardous because their fine particles can get into the eyes and digest sensitive tissues, causing blindness. Exercise extreme caution when weighing and otherwise handling these enzymes.
15. Always use a mechanical device for pipeting corrosive, radioactive, toxic, or highly infectious liquids. In fact, it is safe practice never to pipet anything by mouth.

Chemical Disposal

1. Follow all directions for disposing of residues and unused portions of reagents.
2. No excess chemicals of any nature should be thrown into refuse receptacles. They should be properly disposed or destroyed; or the safety officer should be called on to dispose of them.
3. Prepare a complete list of chemicals of which you wish to dispose.

4. Classify each of the chemicals on this disposal list as a hazardous or nonhazardous waste chemical. (Check with the local environmental agency office for details.)
5. Unlabeled bottles (a special problem) must be identified to the extent that they can then be classified as hazardous or nonhazardous wastes and disposed of properly.
6. Allow for materials to cool before disposal. Make sure appropriate disposal techniques are used.
7. Waste materials should be properly stored until their removal from an institution.

Chemical Substitutions for Safety

1. Reduce risks by diluting substances instead of using concentrates.
2. Use micro/semimicro techniques instead of macrotechniques.
3. Undertake all substitutions with extreme caution.
4. Use films, videotapes, computer programs, and other methods rather than experiments involving hazardous substances.

SAFETY COURSES

1. All science teachers should complete a science safety course.
2. Other chemical safety training courses are offered by commercial organizations, universities, professional societies, and trade associations.
3. All science teachers and laboratory workers should complete a first-aid and CPR course offered by the American Red Cross.

IN CASE OF AN ACCIDENT

1. Report injuries promptly to your supervisor.
2. Report to a physician any unexplained disorders such as headaches or skin rash. They may be caused by the continual use of chemicals, especially if certain "bad actor" chemicals are handled without the full protection of an exhaust hood.
3. Report unsafe conditions or hazardous work habits existing in your own area to your supervisor. Report unsafe conditions or hazardous work habits existing outside of your own work area to the department head or the safety officer.
4. Observe all KEEP OUT signs. Remember that personnel who work directly with hazardous materials are guarded by means of x-ray examinations, immunization shots, film badges, etc.; therefore, it is doubly important for all service workers to heed warning signs.
5. A stumble or slip may be much more serious in a chemical laboratory than elsewhere, not only because of the chemical a person may be carrying, but also because it is almost certain to be in a glass container. Do not leave anything on the floor likely to impede the free passage of others, and if you spill grease or liquid, see that it is wiped up immediately.
6. Tetanus toxoid inoculation as a precaution against possible tetanus infection is much superior to a tetanus antitoxin shot after an injury has occurred. You are encouraged to request this tetanus immunization.

7
BIOLOGICAL HAZARDS

INTRODUCTION

As early as 1893, cases of laboratory-acquired infections occurring in research labs were recorded. Cases still routinely appear today in the scientific literature. Microorganisms, like toxic chemicals, are a definite hazard to persons performing biologic experiments. Working with them requires particular handling techniques and possibly specialized laboratory equipment. Instructors must be aware of the hazard posed by infectious agents and the possible sources of infection present in the laboratory. Biological experimentation offers a unique and interesting learning experience for the student, and when properly conducted, it can be very informative and exciting.

ACCIDENTAL INFECTIONS

In approximately 80% of all laboratory-acquired infections, the cause is unknown. Such a minor mishap is required to release pathogenic microorganisms into the air that pinning down the exact time and operation responsible may be impossible. In the 20% of laboratory infection cases for which the causes are known, there are five most frequent causes.

1. Oral aspiration through pipettes
2. Accidental syringe inoculation
3. Animal bites
4. Spray from syringes
5. Centrifuge accidents

Other common causes of laboratory-acquired infections are cuts or scratches from contaminated glassware, cuts from animal autopsy instruments, and the spilling or dropping of pathogenic cultures on floors or table tops. Table 7-1 lists the various specimens and samples that must be assumed to contain infectious agents.

Laboratory aerosols that enter the body through the respiratory tract are known to be sources of infection. Spray from syringes, centrifuge accidents, a film of culture breaking on an inoculating loop, and a surface bubble breaking when a culture is stirred all give rise to aerosols that can readily enter the body.

Another common source of infection by microorganisms is contact with laboratory animals. Laboratory animals transmit pathogenic microorganisms to humans through simple contact, bites, scratches, generation of aerosols, and contact with a contaminated cage or bedding.

This information is presented to form a basis for proper handling of pathogenic organisms. It is strongly recommended that pathogenic organisms *not be introduced* into a school science laboratory experiment. Only sterile biologic fluids and nonpathogenic microorganisms from reputable scientific supply houses should be used.

If laboratory animals are to be used, they should be obtained through licensed experimental animal suppliers and should be housed and cared for so that they will not acquire infection during experimentation.

TABLE 7-1. Specimens and Samples Presumed to Contain Etiologic Agents

Blood and blood fractions
Sputum
Urine
Body fluids, all types
Tissues
human
primates
all other
Cultures (nose, throat, cervix, lesions, etc.)
Sewage samples
untreated
disinfected
Environmental
water samples
soil samples

Infectious Hazards

Hazards in the biological research laboratory are principally infectious. They are difficult to combat compared with chemical, radiological, mechanical, electrical, and fire hazards.

The reasons for this are several:

1. The disease is more difficult to detect and assign as occupationally acquired.
2. Even if the disease is determined to be occupational, less than one-third of all cases can be traced to a definite act or accident.
3. Evaluated information, rules, regulations, codes, and standards relative to research hazards and preventive measures often are not available.
4. A systematic "job analysis" of the project relative to safety is not a conscious part of the research plan.
5. Medical and biological personnel as a rule tend to be more reluctant than, for instance, engineers or chemists to enter into a professionally planned safety program that involves critical scrutiny of the entire research process.

Microbiological safety in its simplest form relates to the precise control of microbial elements in any particular environment. Its application in laboratories where pathogenic cultures or infected animals are being used will help to prevent infections in laboratory workers.

A second reason for microbiological environmental control is to protect the validity of an experiment. In the absence of suitable controls, laboratory results can be confounded by the accidental or unintentional transfer of infectious microorganisms from animal to animal or from test tube to test tube.

A third reason for microbiological environmental control is to protect people from infection by laboratory animals not known to be infected. Human infections may result from any laboratory use of animals. For example, since 1934 there have been 18 reported cases of monkey B virus infection identified with the handling of monkeys or their tissues. Most of these cases were fatal; the causative virus apparently occurs naturally in monkeys without producing ill effects. Another example is the observation of an abnormal incidence of hepatitis among persons handling apparently healthy subhuman primates, chiefly chimpanzees. From 1953 to 1962, some 69 cases were documented in which primate-to-human transfer was suspected. A third potential hazard exists for human handlers who have close contact with monkeys that have acquired tuberculosis.

First, we will review the known and probable causes of laboratory infections. Then we will consider the available methods of prevention, dealing with five approaches to laboratory safety and outlining some specific recommendations for each.

Causes of Laboratory Infections

The true causes of accidents lie in a combination of circumstances rather than the simple, direct effect of one or two external agents. This is why cause analyses can conveniently follow an epidemiological approach, wherein the total interrelationships and interactions of the host, the accident agents, and the environment are considered.

In the infectious disease laboratory, there are a number of ways in which the major elements present interact with each other. The animal host or person affects and is affected by the environment, physical agents, animals, and even insects in various ways. Of course, it is not the interactions themselves that are responsible for accidents; some interactions are normally required for carrying out laboratory functions. But accidents do happen when the sequence is wrong, when timing is incorrect, when the amount used is too much or too little, when the wrong choice is made, or when there is a combination of any or all of these factors.

We have already noted that naturally infected monkeys and chimpanzees can transmit diseases to humans in the laboratory. There are, in fact, well over 100 diseases of animals that could conceivably be passed to man. About a dozen of these zoonoses are known to have been transferred from naturally infected animals to man in the laboratory: lymphocytic choriomeningitis, infectious hepatitis, cat scratch fever, Newcastle disease, psittacosis, monkey B virus, leptospirosis, tuberculosis, malaria, amebidsis, shigellosis, and streptococcal and staphylococcal infections.

Because naturally infected animals can cause laboratory infections, we may also expect animals challenged in the laboratory with infectious disease agents to be capable of transmitting infection to humans. Although there are no definitive data showing the frequency with which this happens, we do know that the hazard exists. Moreover, a sizable amount of supportive evidence has been provided by animal cross-infection studies. In these studies, control animals are usually caged together with infected animals or in adjoining cages. Then periodic tests are made to determine if the controls themselves become infected. The frequency with which cross-infection of this type occurs provides good presumptive evidence of the hazards that may exist for man in the same environment.

Infections from animals may be caused by bites or scratches, from contact with contaminated cage debris, or from breathing of airborne organisms from sick or coughing animals. If animals have been used in aerosol experiments, infectious organisms on the fur may be released into the air, thus possibly infecting animal attendants.

The specific primary causes of laboratory infections fall into two groups. One group, accounting for about 20% of the total, consists of recognized accidents that result in the infection of laboratory personnel. The second category, which includes the other 80%, consists of accidents whose causes are often classified as "unknown" because there were no previously recognized or recorded accidents or incidents that could be shown to have been responsible for the infections. Although this is a somewhat unique situation, that it is true is shown by the fact that the percentage of unknown causes has remained reasonably consistent in the various infection surveys.

Because unsafe acts or unsafe conditions have not been identified for approximately 80% of the recorded laboratory infections, some laboratory procedures and equipment have been suspected of creating hazards. Indeed, this suspicion has been confirmed by a number of studies in which the amount of microbial aerosol produced by various laboratory techniques was measured. Most of the usual techniques have been tested in the United States and England. These studies show that most common laboratory techniques carried out in the ordinary manner will produce infectious airborne particulates. At least one study has indicated that these particulates are of a size that will readily penetrate the human lung if they are breathed. Of course, these results only suggest possible means of laboratory infection. The type of microorganism, its probable infectious dose, its environmental resistance, the resistance of the host, and many other factors would have to be evaluated to accurately define a hazard.

Laboratory infections caused by accidents that can be identified with unsafe acts and conditions or with procedures and techniques that unsuspectingly release infectious aerosols into the laboratory environment illustrate the fact that laboratory safety is a problem of environmental control. The microbe must remain in its environment (test tube, flasks, etc.) and the microbiologist must be protected from the organism's environment. Although this solution appears simple and straightforward, its application is complex. Microbes capable of causing human infection are not readily detectable in the usual sense; the infecting dose may be odorless, tasteless, and invisible to the eye.

Statistical studies of accidents and infections at several large laboratory institutions provide some additional epidemiological data of probable significance in infection prevention. These are summarized as follows:

1. In general, more infections are associated with manipulating cultures than with handling animals.
2. Those most closely associated with the infectious operations, laboratory technicians, students, trained professional personnel, and animal handlers, are in the greatest danger of becoming infected.
3. Among those who work directly with pathogens, it is younger persons with less formal training who are more apt to become infected.
4. Inhalation of infectious aerosols is by far the most frequent mode of laboratory infection.
5. The physical form of an infectious agent is related to its hazard level. Dried or lyophilized cultures and infected eggs are more dangerous to handle than liquid cultures or infected blood specimens.

Infectious Agents

Nearly all groups of microorganisms have some effect on man. Table 7-2 presents the various groups of microorganisms and some of the diseases in man for which each group is responsible.

Each group listed is responsible for many more diseases than the few listed. There is no group of mi-

TABLE 7-2. Microorganisms and the Diseases They Cause

Microorganisms	Human Disease
Bacteria	diphtheria
	tuberculosis
	rheumatic fever
	pneumonia
Viruses	chicken pox
	measles
	mumps
	poliomyelitis
Fungi	athlete's foot
	systemic mycosis
Rickettsiae	typhus
	Q fever
	Rocky Mountain spotted fever
Protozoa	Schistosomiasis
	malaria
	Giardiasis

Source: NIOSH, *Safety in the School Science Laboratory,* Instructors Resource Guide, 1979, p. 10–22.

croorganisms that does not contain some pathogenic members. Consequently, experimentation involving microorganisms either directly or indirectly must be strictly controlled.

Animal-to-Man Diseases (Zoonoses). Laboratory animal illnesses are important because of their potential threat to the health of animal laboratory personnel. There are more than 150 diseases infectious from animal to man. Controlling such diseases is the most important step in the protection of the employee and student.

Outlined below are a few examples of diseases infectious from animal to man.

1. Viral
   (a) *Viral encephalitis:* Occurs in mice, birds, guinea pigs, dogs, monkeys, rats, pigs, and horses. Transmitted by direct contact with the conjunctive, respiratory, or digestive tract, or urine of mice. Influenza-type illnesses, meningeal symptoms can be fatal.
   (b) *Rabies:* Occurs in dogs, cats, foxes, bats, skunks, and monkeys. Enters the body via virus-laden saliva entering a wound caused by the bite of a rabid animal, or if infected saliva comes into contact with fresh open wounds or the mucous membranes of the eye, nose, or mouth. It can also be contracted by inhaling the virus-laden air found in caves housing bat colonies. Marked by headache, weakness, elevation of temperature, nervousness, anxiety, convulsions, paralysis, and death.
   (c) *Monkey B virus:* Monkeys. It is a residue virus in monkeys that usually gives rise to a mild disease of short duration or produces encephalomyelitis in monkeys. Appears as lesions on the tongue and/or lips. Transmitted when the victim is bitten or scratched by an infected monkey. Affects the brain, spinal cord, and spleen and is almost invariably fatal to man.
   (d) *Ornithosis (psittacosis):* Found in parakeets, poultry and quail, parrots, and pigeons. In man, it is usually contracted by inhaling air contaminated by droppings from infected birds. Produces severe respiratory-related illness.
2. Rickettsial
   (a) *Q-fever:* Present in cattle, sheep, goats, poultry, and rodents. Transmitted via milk or inhalation of contaminated air. Symptoms are similar to those associated with a bad case of influenza.
   (b) *Rocky Mountain spotted fever:* Present in dogs and rabbits. Transmitted by ticks. Symptoms are skin rash, headache, and fever. Involves the liver, kidneys, and lungs; can be fatal.
3. Bacterial
   (a) *Tuberculosis:* Found in monkeys and cattle. Transmitted by inhalation of spores in contaminated air or direct contact with saliva. Affects the lungs.
   (b) *Brucellosis:* Occurs in cattle, swine, sheep, and goats. Contracted directly from infected animals and/or their tissues or possibly by insects. Causes abscesses of the lungs and, later, bone destruction.
   (c) *Anthrax:* Present in pigs, sheep, goats, and horses. Contracted through contact with an infected animal's skin or hair or inhalation of contaminated air. The disease is characterized by subcutaneous nodules and abscesses of the lungs, intestines, and brain.
   (d) *Tularemia:* Present in rabbits, squirrels, and rodents. Transmitted by bites of ticks, fleas, and lice. Affects the lymph nodes and lungs.
   (e) *Tetanus:* Found in rodents, rabbits, and dogs. Transmitted by animal bites or contact through open skin. Affects muscles (lockjaw).
   (f) *Salmonellosis:* Found in almost every domestic and even wild animals. Contracted from contaminated food or drinking water or inhalation of infected dust. Causes severe gastroenteritis or even death.
4. Protozoan
   (a) *African sleeping sickness:* Found in the blood stream of wild and domestic animals. Transmitted to man via the tsetse fly. The organism multiplies in the blood stream and liberates byproducts that affect the metabolism and produce a fever. If not treated promptly, the organisms invade the cerebrospinal fluid and, ultimately, cause death.
   (b) *Chagas' disease:* Found in dogs, cats, and a wide variety of wild mammals. Transmitted to man via a Reduvid bug. May cause fatal myocarditis, severe megacolon, or megaesophagus.
   (c) *Malaria:* Widespread in birds and mammals. Transmitted to man by the anopheles mosquito. Characterized by fever, enlarged spleen, anemia, and even death.
   (d) *Amoebic dysentery:* Found in monkeys. Usually contracted directly from contaminated food or water. Causes infection of the intestinal mucosa, diarrhea with loss of blood, and even death.
5. Helminths (Parasitic Worms)
   (a) *Trematodes (flukes):* Found in the blood, liver, and muscles of rodents, cats, dogs, sheep, fish, and birds. May be obtained by eating contaminated food or infected water. Usually in-

volves an intermediate host. Causes damage to the liver, skin, lungs, heart, and sometimes death.

(b) *Cestodes (tapeworm):* Found in swine, rodents, dogs, and cats. Transmitted through contaminated food or via an intermediate insect host. Usually involves the muscles and intestinal tract, but can invade the central nervous system and cause death.

(c) *Nematodes:*

i. *Whipworm:* Commonly found in rodents, swine, and dogs. Lives in the liver and intestines. Eggs are released through feces or upon death of host and infection develops by ingestion of the eggs. Characterized by blood in stools, abdominal discomfort, or prolapsed rectum.

ii. *Trichina:* Infects many animals, but especially swine, dogs, cats, and rats. Lives in the intestine, but penetrates the digestive tract and may be found in practically every organ, although muscles are the organs of choice. Man usually is infected by eating raw pork. Characterized by weakness, muscular aches, and twitching. Can be fatal.

iii. *Roundworm (ascaris):* Lives in the small intestine of swine. Eggs are eliminated through the feces. Larva are eaten by man, migrate to the liver and lungs, are coughed up, swallowed, and grow to adult size in the small intestine. May damage the liver and lungs during migration, and large numbers in the intestine can lead to blockage.

iv. *Hookworm:* Found in dogs and cats. Reside in the small intestine. Transmitted by larva through skin penetration. Problems occur with migration from the skin to the lungs, intestines, etc. Characterized by enlarged heart, muscle tenderness, abdominal cramps, hemorrhage, and even death.

v. *Pinworm:* Live in the cecum and intestine of rodents, cats, and dogs. Eggs are discharged with the feces and ingested by man. Causes inflammation of the intestine or appendicitis.

6. *Arthropods:* Includes ticks, mange mites, lice, fleas, and bedbugs.

Prevention of Laboratory Infections

As with any type of preventive effort, the prevention of laboratory-acquired infections should begin with an assessment of the extent of the problem or an estimate of the probable extent of future problems. Potential hazards may result from research being carried on with infectious cultures, from the use of cultures or infected tissues in classroom demonstrations, from clinical diagnostic procedures, or from the use of animals in laboratory situations. In any case, there must be some understanding of and agreement on the dangers by administrators and laboratory directors. Even when the microbiological hazards are understood, the philosophy of scientific freedom that characterizes academic life can often work to oppose the inspections, investigations, and regulatory requirements of a good safety program.

Once an adequate assessment of the potential microbiological hazards is made and management commits itself to a preventive program, a precise personnel policy regarding occupational health should evolve. Management should then make a series of policy decisions related to the goals of the safety program and how it is to operate. A list of policy questions is provided for those concerned with the construction of laboratory facilities for infectious disease work. Many of these questions apply to the safety policy as a whole.

1. What level of occupational infection is acceptable to management? Is it desired to prevent all work-incurred infections, including subclinical infections, that can be detected only serologically? Or is management's aim to prevent only those infections that are likely to result in incapacitating illnesses, or only those for which there is no treatment.
2. To what extent is the control of microbiological hazards to be extended to protect persons in areas peripheral to the laboratory? Public relations, economic and legal considerations are involved here.
3. What type of supporting medical program is to be provided for persons at risk in the laboratory.

A program for controlling microbiological hazards should begin with a clear concept of the goals, an understanding of the nature of the hazards involved, and an expression of the policies to be followed in achieving control. By this action, an administration establishes responsibility for safety control, includes planning for accident control in all phases of laboratory work, and makes it clear that no job will be considered so important that it cannot be done safely.

Primarily, the cardinal points of microbiological control will be education, engineering, and enforcement. The detailed implementation of safety control can be discussed by considering five important elements. Each element's use is determined by the extent of microbiological hazards and by management's policy concerning them. Management aspects, vaccination, and safe techniques and procedures will be discussed here.

Management Aspects

Some of the programming and policy responsibilities have already been discussed. Management at various levels must also concern itself with the proper selection of laboratory employees. Proper selection refers not only to technical competence, but also to the fact that it may be undesirable to employ persons with certain physical conditions for work with some types of infectious agents for medical reasons.

Management should likewise be concerned with providing safety training for laboratory personnel, formulating safety regulations, and establishing methods for adequately reporting and investigating accidents.

Management should also attempt to control the human factors that result in accidents and strive to provide an atmosphere in which personnel may develop attitudes conducive to safe performance. Practical experiences have shown that such an approach is essential for an accident and infection prevention program to be successful. Good laboratory management includes good safety management since safety is an essential part of any productive enterprise.

Vaccination

The vaccination of laboratory personnel is recommended when a satisfactory immunogenic preparation is available. A doctor should be consulted on available vaccines.

The efficiency of vaccines for laboratory workers is generally evaluated on the basis of their effectiveness in preventing disease in the general population, but the laboratory worker may be exposed to infectious microorganisms at a higher dose level than the public, and such exposure may be by a route different from that normally expected; for example, respiratory infection with the tularemia or anthrax organism.

Education in Microbiological Safety

In spite of the considerations, approaches, and equipment discussed thus far, human factors in accident prevention occupy a dominant position and the educational process is therefore essential in safety.

Endowing the student with technical knowledge is not enough. He or she must be taught how to use the instruments and apparatus of the laboratory. The student must, in the learning process, be made to understand the importance of safety-related laboratory procedures, and be impressed with the notion that a good scientist is also a safe scientist. Too often, school authorities have avoided the need for microbiological safety education and solved infectious hazards problems simply by forbidding the use of pathogenic agents in the school's laboratories.

It is true that not all microbiologists handle or need to handle infectious organisms in their work. But it is a form of "buck-passing" on the part of academia to tell a student taking a course in infectious diseases that if she is required to handle pathogens later in her career, her employer or someone else will give her the proper instructions on their use. If microbiologists and others who work with infectious microorganisms are to be given every opportunity to protect themselves from acquiring occupational diseases, should not safety education in the hazards associated with handling highly virulent microbes be included in a college curriculum? If safe behavior is a requirement and important attitudes are formed early, is it realistic to wait until after the completion of professional training to institute education in safety?

As has been noted, the use of animals and infectious agents often leads to occupational disease among laboratory people. From the point of view of each safety administrator, it is important that an evaluation be made of actual, potential, or future microbiological hazards before deciding if and how much of a prevention effort is required.

It has been principally over the last two decades that attention has been given to the problem of correcting or reducing laboratory-acquired illnesses. Former traditions of personal sacrifice are gradually becoming outdated by economic, moral, and legal pressures. Also, in the last few years, it has become eminently clear that laboratory determinations will be accurate only if controlled to the extent that concurrent culture cross-contamination or animal cross-infection can be prevented. This has prompted research helpful in developing techniques and methods that reduce human infectious risks in the laboratory. The most important single conclusion from this research is that preventing the release of accidental microbial aerosols at the laboratory working surface through careful techniques and through the use of containment devices is the best way of achieving microbiological environmental control.

The specific tools for controlling microbiological hazards are the following:

1. Management and administrative tools: reporting, analysis, regulation, and training
2. The use of correct techniques
3. The use of safety equipment

4. Properly designed laboratories
5. Vaccination of personnel

Evaluation of Microbiological Hazards

Like other aerosols, microorganisms can be collected from the air through the use of a number of air-sampling procedures. Microorganisms can be collected by impingement in liquids, impaction on solid surfaces, filtration, sedimentation, centrifugation, and electrostatic precipitation. They must then be cultured in traditional fashion for identification and quantification.

Conclusion and Summary

Each person in a laboratory is responsible for preventing accidents and infections during the course of his individual actions. All supervisors should train persons under them in safe working habits.

No job should be considered so important that it need not be done safely. Preplanning for safety should be part of all programs. This should include procedures to be followed in the event of spills, and in the case of especially hazardous procedures, pilot runs should be made using simulations. Surface or air samples can be taken during these procedures.

All injuries or illnesses should be immediately checked by competent medical authorities. One of the reasons for this is that so many illnesses resemble those studied in the laboratory that occupationally acquired infections are difficult to detect. Whenever possible, laboratory workers should be immunized.

Whenever possible, hazardous operations should be carried out in biological safety cabinets. These should be ventilated, but often nonventilated glove boxes can also be used. When using gas in safety cabinets, you should have at least 50 cfm airflow through the cabinet for each gas outlet. All valves in the cabinet should be labeled.

All infectious or toxic materials, equipment, or apparatus should be autoclaved or otherwise sterilized before being washed or disposed of. Special precautions should be taken with materials that are disposed through the sewage system.

The researcher is also responsible for any maintenance personnel working in the laboratory or on its utilities.

The same precautions that apply to human workers also apply to preventing cross-infections between experimental animals. Any organisms released by exposed or infected animals present a hazard to experimenters, animal caretakers, and other experimental animals. A study of the literature shows that many organisms present such a hazard.

PROPER LABORATORY PROCEDURES

Laboratories can be classified as follows:

Routine. Laboratory restricted to the handling of organisms that do not present any special hazards to trained laboratory workers when handled with conventional microbiology laboratory techniques, for example, coliform organisms, E. coli, fecal streptococci, Clostridium perfringens, nonselective agar plate counts, bacteriophage, fluorescent pseudomonads, autotrophic bacteria such as iron bacteria, sulfate-reducing bacteria, nonpathogenic yeasts, molds, actinomycetes, and other nonfecal indigenous water bacteria.

General Pathogen. Laboratories approved as suitable for the handling of pathogenic organisms presenting less hazard to trained and qualified laboratory workers than those in category B. Such organisms include Salmonella species (other than Salmonella typhi and Salmonella paratyphi A), Shigella (other than Shigella dysenteriae), Campylobacter, Listeria and Yersinia (other than Yersinia pestis) together with enteric viruses. Such laboratories may use accepted techniques that might isolate category B pathogens should these be unexpectedly present in the original sample, providing any such organisms are immediately destroyed or transferred at once to a category B pathogen laboratory as soon as their isolation is suspected.

Category B Pathogen. Laboratories approved as satisfying the requirements for handling category B pathogens (see Table 7–3 for a list of organisms).

Approval of Laboratory Facilities for Particular Work

All bacteriology laboratories must be classified and approved by the director of the department or equivalent, and/or the regional medical adviser or equivalent after consultation with the principal microbiologists.

Before handling any potentially pathogenic organism not specifically mentioned in this chapter, appropriate advice as to the class of laboratory facilities necessary must be obtained.

No laboratory should knowingly handle or seek to isolate organisms that it is not approved to handle. It is recognized that many acceptable procedures carry some risk of isolating such organisms (e.g., an agar plate count in a routine laboratory may include Salmonella colonies), and in the event a laboratory suspects the isolation of organisms for which it is not approved, all such isolates must be immediately destroyed by an approved method using the necessary

TABLE 7-3. Category B Pathogen Classification

Arboviruses (except Semliki Forest, Uganda S, Langat, yellow fever 17D vaccine strain, and Sindbis viruses)
Bacillus anthracis
Bartonella spp.
Blastomyces dermatitidus
Brucella spp.
Chlamydia psittaci (excluding C. psittaci GPIC, EAE, and meningopneumonitis strains)
Clostridium botulinum
Coccidioides immitis
Coxiella burneti and other pathogenic rickettsiae
Cryptococcus neoformans
Francisella tularensis
Histoplasma capsulatum and related species
Legionella pneumophila and related organisms
Leptospira species
Mycobacterium tuberculosis and other species of pathogenic mycobacteria
Paracoccidioides brasiliensis
Pathogenic amoebae
Pseudomonas mallei (the glanders bacillus)
Pseudomonas pseudomallei
Salmonella paratyphi A
Salmonella typhi
Shigella dysenteriae
Vibrio cholerae
Yersinia pestis (the plague bacillus)

Source: National Joint Health and Safety Committee for the Water Service, *Safety in Microbiology Laboratories in the Water Industry,* London, 1983.

precautions or transferred with due care to an appropriately approved laboratory.

The hazards of microbial infection are not the same in all laboratories, and the procedures and facilities that are necessary for laboratories handling pathogenic bacteria would be unnecessarily restrictive for laboratories handling only more innocuous microorganisms. It must be borne in mind, however, that many so-called "nonpathogenic" organisms can under certain circumstances cause infection, and furthermore, media designed for the growth and recognition of organisms such as coliforms or E. coli may also permit the growth of pathogens should any be present. It is, therefore, important to observe certain minimum standards in all bacteriological laboratories.

General Precautions

Requirements of a routine laboratory include the following.

Structural. The laboratory should not be part of the general traffic pattern, and it should be possible to restrict access to persons directly involved with the work of the laboratory. Floors should be of a suitable impervious nonslip surface. Laboratory doors should be equipped with suitable view panels, and equipment must be located to preclude any accidents caused by opening doors. An area outside the laboratory must be provided for the consumption of food and drink.

Furnishings. Laboratory bench surfaces must be readily cleanable, impermeable, and resistant to disinfectants. There should be adequate storage space within the laboratory for all immediate requirements; it should not be necessary to store items on the floor, under benches, or on top of cupboards. Facilities must be available for the storage of laboratory coats within the laboratory, and everyday outer clothing outside the laboratory.

Equipment and Services. The laboratory should have sufficient power outlets to supply all equipment without recourse to multiple adaptors and sufficient suitably located outlets for other necessary main services. Adequate facilities must be available for the washing and drying of hands before leaving the laboratory (wherever possible, a separate wash basin should be provided). Sufficient incubation space and adequate autoclaving facilities must be readily accessible and a sufficient supply of suitable disinfectant and discard jars must be provided. Stocks of reusable equipment must be sufficient for the laboratory to operate in accordance with safe practices, and adequate facilities for the cleaning of such equipment must be available.

Additional Requirements for Pathogen Laboratories. In addition to the requirements above, pathogen laboratories must be physically separate enclosed areas, entered through a door that must be locked when the laboratory is unoccupied. The door must be clearly marked with a biohazard label and a warning to unauthorized persons not to enter.

There must be a wash basin near the door, fitted with wrist- or foot-operated taps and a hot-air hand dryer, and a continuous-flow roller towel. Otherwise, paper towels must be provided and adequately maintained.

Any actively induced ventilation must be in the form of powered air extraction directly discharging to the open air and located so as to prevent the discharged air from flowing into adjacent rooms or buildings.

Safe Techniques and Procedures

The same general precautions required in any science laboratory using hazardous chemical compounds are required in any laboratory in which microorganisms are studied or used. Sound fundamental laboratory techniques, well supervised and conscientiously carried out, can do much to achieve environmental control and reduce the hazards of infection. Many procedural rules are obvious because their aim is to prevent direct contact with harmful microbes. Others may be less well understood because their purpose is to prevent airborne contamination of the workers' environment at a level where such contamination is not easily or readily detected. Infectious aerosols are like dangerous radiations, except that the former are more difficult to monitor. A list of procedural rules that are widely applicable in infectious disease laboratories follows:

1. Never directly pipet by mouth infectious or toxic fluids; use a pipetor.
2. Plug pipets with cotton, even though used with a pipetor. This minimizes contamination of the pipetor.
3. Do not blow infectious material out of pipets.
4. Do not prepare mixtures of infectious material by bubbling expiratory air through the liquid with a pipet.
5. Use an alcohol-moistened pledget around the stopper and needle when removing a syringe and needle from a rubber-stoppered vaccine bottle.
6. Use only needle-locking hypodermic syringes. Avoid using syringes whenever possible.
7. Expel excess fluid and bubbles from a syringe vertically into a cotton pledget moistened with disinfectant, or into a small bottle of cotton.
8. Before and after injecting an animal, swab the site of injection with a disinfectant.
9. Sterilize discarded pipettes and syringes in the pan where they were first placed after use.
10. Before centrifuging, inspect tubes for cracks. Inspect the inside of the trunnion cup for rough walls caused by erosion or adhering matter. Carefully remove all bits of glass from the rubber cushion. A germicidal solution added between the tube and the trunnion cup not only disinfects the surfaces of both of these, but also provides an excellent cushion against shocks that otherwise might break the tube.
11. Use centrifuge trunnion cups with screw caps or the equivalent.
12. Avoid decanting centrifuge tubes; if you must do so, afterwards wipe off the outer rim with a disinfectant. Avoid filling the tube to the point that the rim becomes wet with culture.
13. Wrap a lyophilized culture vial with disinfectant-moistened cotton before breaking. Wear gloves.
14. Never leave a discarded tray of infected material unattended.
15. Sterilize all contaminated discarded material.
16. Periodically, clean out deep-freeze and dry-ice chests in which cultures are stored to remove broken ampoules or tubes. Use rubber gloves and respiratory protection during this cleaning.
17. Handle diagnostic serum specimens carrying a risk of infectious hepatitis with rubber gloves.
18. Develop the habit of keeping your hands away from your mouth, nose, eyes, and face. This may prevent self-inoculation.
19. No food or drink should be stored, taken into, or consumed in the laboratory. Smoking should not be permitted in the laboratory.
20. Make special precautionary arrangements for respiratory, oral, intranasal, and intratracheal inoculation by infectious material.
21. Protective clothing should be worn when necessary and should be carefully disposed of. Give preference to operating room gowns that fasten at the back.
22. Evaluate the extent to which hands may become contaminated. With some agents and operations, forceps or rubber gloves are advisable.
23. Wear only clean laboratory clothing in the dining room, library, and other nonlaboratory areas.
24. Shake broth cultures in a manner that avoids wetting the plug or cap.
25. Only authorized employees or students should be allowed in the laboratory.
26. Students and instructors should wash their hands thoroughly before leaving the laboratory.
27. Extraneous items such as books, coats, and umbrellas should be left outside the laboratory.

These general precautions are designed to minimize the possibility of accidental infection of both laboratory personnel and any outsiders who come into contact with them.

Specific Laboratory Procedures

A number of specific laboratory operations that deserve special attention when microorganisms are involved will be covered here.

Bacteriological Techniques

Pipetting. The greatest hazards of pipetting are (1) the production of aerosols, (2) accidental ingestion of fluid, and (3) contamination of the mouthpiece. Hazards (2) and (3) can be remedied by using a pipetting bulb or another device with the same function. The mouth then does not have to be anywhere near the pipet. Oral pipeting is dangerous and must be prohibited: Cotton wool plugs are not a barrier to infection and a pipet can act as a direct vehicle for the transmission of infection from hand to mouth. The use of pipet bulbs is mandatory. Alternatively, automatic pipets with sterile disposable tips may be used. Cotton-plugged pipets are recommended for use with pipetting devices in order to minimize their contamination.

Pipets should be handled carefully to minimize the hazard of aerosol generation. The pipet should never be used to bubble air through a contaminated liquid. Liquid should never be forcefully blown out of the pipet. The pipet should always be discharged, if possible, with the tip below the surface of the receiving liquid. Immediately after use, contaminated pipets should be immersed in a germicidal solution and then autoclaved before reuse.

Syringes. The hazards common to syringe use are (1) accidental inoculation and (2) aerosol production. Accidental inoculation must be carefully guarded against. If animals are to be inoculated, care must be taken to restrain the animals and prevent them from bumping into a syringe.

Only syringes with locking needles should be used for work with pathogenic organisms. If a nonlocking needle should happen to come off a syringe, a very hazardous aerosol can be generated.

Ideally, disposable syringes with a permanently affixed needle should be used. Excess liquid or air bubbles should be expelled vertically from a syringe into a piece of cotton moistened with a disinfectant. A syringe should never be used for mixing liquids by forcefully expelling a liquid from the syringe into another liquid. When a syringe is used to transfer one liquid to another, the tip of the syringe should always be placed below the surface of the receiving liquid before a liquid is discharged. After use, syringes should be placed in a container of disinfectant and then autoclaved.

Students and instructors must take great care when manipulating syringes. The hand should never touch the needle and hub or the shaft of the plunger. These areas are often contaminated in normal use.

Inoculating Loops. Inoculating loops must be used with care. The film held by a loop may break and cause substantial atmospheric contamination. Inoculating loops should be kept fairly short, about 2 in. to minimize the possible production of aerosols by vibration of the loop.

Liquid cultures should never be agitated by inoculating loops because of the possible production of aerosols. Loops should be allowed to cool before insertion into liquids. This may require the use of more than one loop so that as one is being used, others are cooling.

Whenever inoculating loops are used, any actions that might result in the generation of an aerosol—jerky movements, shaking of the loop, agitating liquids—must be avoided.

Loops must be flamed with care to prevent boiling and volatilization of the material before the flame can kill all pathogenic microorganisms. This may be achieved by inserting the loop into the cool center of the Bunsen flame and slowly withdrawing it through the hot part of the flame. Alternatively, a hooded Bunsen or loop incinerator is preferred and for work in any pathogen laboratory is essential. Plastic loops may also be used with safe disposal into disinfectant in a discard jar after use. Such loops must be completely submerged overnight before disposal or else autoclaved in the discard jar.

Aerosols and Droplets. Infection can arise from inhalation of aerosols or of fine droplets of contaminated material. Extreme care must therefore be exercised with many common operations and accidents, including:

1. Operating a screw-cap universal container. Plug-stoppers or snap-on closures should not be used for this reason.
2. Accidental breakages.
3. Breakage of containers in centrifuges.
4. Mechanical homogenizers.
5. Ultrasonic agitation.
6. Bubbling of air through cultures of microorganisms, shaking or homogenization by mechanical means. Cultures should be mixed by rolling bottles between the hands.
7. Slide agglutination readily produces aerosols and droplets. Care must be taken to minimize the extent to which films of liquid are broken

when making suspensions and adding sera, and the use of formol-saline for a suspending medium is recommended.

Handling Cultures

Dehydrated Media. Before handling chemicals or ready-prepared media, all warning labels placed on bottles must be observed. Manufacturer's instructions must always be observed when reconstituting dehydrated media. These are prone to produce fine dusts when being weighed out and mixed, and therefore, care should be taken to minimize this. Suitable and effective respiratory protection should be provided for, and worn by staff preparing media when the risk of inhaling such dusts exists.

Spillages on bench or balance must be avoided as much as possible and the area always thoroughly cleaned after use. Spatulas and glassware used with dangerous chemicals must be decontaminated by thorough rinsing before release for final wash-up. Solid materials should not be added to very hot liquids as they may cause the liquid to erupt.

Agar Media. Agar will only produce a gel in water after heating to 98–100°C. This may be achieved by "flash" autoclaving at a steam pressure of 5–10 psi for a short period of time, heating in free steam and/or boiling while stirring. Glass vessels containing agar media must never be heated by a direct flame. Heat-resistant gloves or similar protection permitting ease of manipulation must always be worn for the safe handling of hot materials, especially molten agar.

The number of inoculated Petri dishes stacked or carried by hand should be restricted to about five. If more than this number are to be handled, then they should be carried in special containers or deep trays. Similar care should be taken with other culture vessels. Racks or baskets of the correct size should be used at all times.

Bench Work

In routine laboratories, all working surfaces should be swabbed down with a suitable disinfectant at the end of each working day.

In general, category B pathogen laboratory surfaces should be disinfected at the end of each sequence of operations. When working with infected material or cultures, it is advisable to work on an absorbent surface such as Benchkote, or absorbent paper tissues moistened with a suitable disinfectant. This helps to contain splashes and spillages.

Breakages and Spillage

The biological safety officer must be informed of any major breakage or spillage, and the incident recorded in an incident book.

Broken cultures or spillage must be attended to immediately, by covering them either with a cloth or paper tissues soaked in disinfectant or disinfectant powder.

After not less than 10 min, the material should be cleared away, wearing disposable gloves and using swabs and a metal dustpan. The waste material should be placed in an infected waste container and autoclaved. The dustpan should also be placed in a suitable container and autoclaved.

Centrifuging

Centrifuges are commonly used in science laboratories to separate cellular material from a suspending liquid medium. If the cellular material consists of pathogenic microorganisms, exceptional care must be taken with all phases of the operation.

Glass tubes used to hold cultures may break in the centrifuge. This may result in large numbers of microorganisms being spread throughout a laboratory. All glass tubes used in the centrifuge should be carefully inspected for cracks and flaws beforehand. One way of minimizing the effects of tube breakage is to fill the space between the glass tube and metal cup with a germicidal solution. If a tube does break, the germicidal solution will tend to nullify the effects of the breakage.

See Chap. 5 for a procedure to clean up broken centrifuge tubes containing bacterial cultures.

Handling Embryonated Eggs

Eggs that are infected with viruses can be extremely dangerous. The inoculation procedure itself can generate a hazardous aerosol in many ways. The infected egg contains an extremely concentrated virus population. The egg shell provides little protection and is readily breached. Infected eggs should be handled only in a ventilated safety cabinet.

Safety Cabinets

Microbiological safety cabinets are specially designed fume hoods for use with pathogenic microorganisms. They may be designed with a small open work area or with a totally enclosed work area. Air velocity across the open face should be greater than or equal to 200 ft per minute. All air exhausted through these cabinets must pass through an absolute filter before it is emitted to the atmosphere. It is recommended that all operations involving pathogenic organisms be performed in a safety cabinet.

As for safety cabinets in category B pathogen laboratories, only class 1 (open-fronted total exhaust) safety cabinets should be used and they must be correctly sited, and regularly tested and maintained.

Signs

The Occupational Safety and Health Administration (OSHA) of the U.S. Department of Labor in 29 CFR, Part 1910.145 requires the posting of a biological hazard warning sign to "signify the actual or potential presence of a biohazard and to identify equipment, containers, rooms, materials, experimental animals, or combinations thereof, which contain or are contaminated with, viable hazardous agents."

The term "biohazard" is defined as meaning only those infectious agents that present a risk or potential risk to the well-being of man.

Figure 7–1 presents the recommended biohazard warning sign. The symbol design must be colored a fluorescent orange or orange-red. The background color is optional as long as it provides sufficient contrast to the symbol.

CAUTION

FIGURE 7–1. Biological-hazard (biohazard) symbol. (From the Code of Federal Regulations Chapter 29, Part 1910.145.)

Blood-Letting Experiments

Experiments involving the observation of human blood cells or blood typing necessarily require a source of human blood. Blood-letting experiments can be very safely conducted if the following rules are observed:

1. Conduct the blood-letting experiment in a neat, clean, and sanitary facility.
2. Only the instructor should puncture a student's finger.
3. Clean the area to be punctured with soap and water and rinse well. Then clean with 70% isopropyl alcohol and dry with sterile cotton or gauze pads.
4. Puncture the finger with a sterile, disposable lancet. Dispose of the lancet properly after its use. Do not reuse the lancet.
5. Hold the finger over a clean microscope slide and let the blood drip onto the slide, making sure there is no contact between the two.
6. Clean the finger with alcohol again once the sample has been collected.
7. Hold sterile gauze or cotton on the area until the bleeding stops, about 3–5 min.

If these few suggestions are followed, a safe and healthy lab experiment can be conducted.

Disinfectants

There is no ideal disinfectant, and selection must therefore be based on the specific situation in which it is to be used. Always wear disposable gloves when handling any disinfectant. The main types of disinfectant suitable for use in microbiology laboratories are the following.

Chlorine Compounds. Examples of these compounds include Chloros or Domestos. They are comparatively cheap and effective against viruses, but are readily inactivated by organic matter and may have poor wetting powers. However, they may be used in conjunction with ionic and nonionic detergents with which they are compatible. They are relatively unstable, especially when diluted, and must be frequently renewed (a hypochlorite solution that does not blacken starch iodide paper should be regarded as ineffective). They are corrosive and therefore should not be used in contact with metals.

They may be used in discard jars and for surface disinfection, preferably in conjunction with a suitable detergent. Chlorine is available in a granular form with the use of chlorinated isocyanuric acid derivatives. These release chlorine on demand and are convenient for dealing with gross spillages and breakages. Examples include Fichlor.

Clear Phenolics. Examples include Hycolin, Sudol, and Clearsol. Phenolic disinfectants are not as cheap as hypochlorites, but they are not readily inactivated by organic material. They may cause the deterioration of rubber. They can be used in discard jars and for surface disinfection.

Formaldehyde. This should only be used for special purposes including the disinfection of safety cabinets or centrifuges following breakages. It produces a highly irritant disinfecting vapor (TLV 1 mg/m^3) that is probably carcinogenic.

Working Concentrations. These should be freshly prepared at frequent intervals, which in the case of hypochlorites must be daily. Commercial preparations of hypochlorite usually contain approximately 100,000 mg/L of available chlorine and the following working concentrations are recommended:

General use: 1% concentration (approximately 1000 mg/L available chlorine).

For discard jars: 2.5% concentration (approximately 2500 mg/L available chlorine).

For heavy contamination (e.g., blood or culture spillage): 10% concentration (10,000 mg/L available chlorine).

Phenolic disinfectants should be used according to the manufacturer's instructions.

Disposal of Infected Material and Cultures

Sterilization of Materials. Sterilization and disinfection procedures are recommended even with nonpathogenic organisms. Heat is considered to be the most effective sterilizing agent. Some gases can be used for sterilization purposes, but they require careful handling and may present dangers themselves. Liquid disinfectants or germicidal agents generally have limited effectiveness and should not be relied on for complete sterilization. No infected material should be discarded without prior sterilization.

Disposable Items. Disposable items such as plastic Petri dishes should be separately discarded in an autoclavable bag held in a suitable leak-proof container, preferably fitted with a loose lid. This container should be clearly marked to indicate the nature of its contents.

Nondisposable Items. These should be set aside in a clearly marked tray or container with leak-proof sides and bottom. Clothing from any pathogen laboratory must be placed in autoclavable bags and autoclaved before release for laundering.

Autoclaving. Batches of material for autoclaving should be identified by means of an autoclave test tape placed conspicuously across the mouth of the container. This tape indicates by a color change whether the container and its load has or has not been autoclaved. A "safe" reading as suggested by tape is not sufficient evidence that autoclaved material is necessarily safe to handle. Browne's tubes or other sterilization indicators may be used similarly and have an advantage in that they may be incorporated within a load. However, proof of safety only follows exhaustive tests under full-load conditions using thermocouples incorporated in the load or spore test strips.

Washing Up. Once autoclaved, nondisposable items should be placed in an area reserved for the accumulation of items to be washed. This should be clearly marked and well separated from the area reserved for items to be autoclaved. Wash-up personnel should wear protective gloves with the prolonged use of hot water and detergent cleaners.

Discard Jars

For used slides, pipets, dispenser tips, disposable loops, spreaders, etc., discard jars of sufficient capacity to permit complete immersion within a suitable disinfectant should be used. Separate discard jars should be employed for disposable and reusable equipment. Discard jars must be left at least overnight before disposal of their contents. They should be emptied frequently and washed out and refilled with the correct strength of a suitable disinfectant. In the case of hypochlorites, this must be daily. Discard jars should never be "topped up" with extra disinfectant, but always refilled. A suitable disinfectant/detergent solution may be used in discard jars to provide cleaning along with disinfection.

Personal Safety

It is the duty of every laboratory worker to protect his health and safety and that of other persons who may be affected by his actions or oversights in the lab. Each person is responsible for his own actions and well-being, but he should also view every operation in light of its possible danger to fellow workers. Personnel must develop an understanding of the nature of the material they are handling and must not deviate from, nor modify, the technical procedures prescribed for their safety.

Protective Clothing. Outer clothing must be left outside the laboratory and laboratory coats worn at all times within the laboratory. Double-fronted, high-collared, tight-sleeved coats are recommended and for category B pathogen laboratory work are essential. They must be fastened; an open coat does not protect against the contamination of everyday clothing. Laboratory coats must be made available for visitors. Laboratory coats must be removed before visiting a rest room, canteen, library, or office. They must be left in the laboratory on the pegs provided and not placed in personal lockers with normal clothing.

A sufficient number of clean coats must be available to allow each worker at least two changes per week. Coats known to be contaminated must be changed immediately and autoclaved before leaving the laboratory, as must all coats used in pathogen laboratories. Laboratory coats should be changed at least once each week in routine laboratories and twice each week in other kinds of laboratories.

Disposable protective gloves should be available at all times for use in operations involving pathogens, blood, sera, or potentially harmful chemical reagents including the oxidase reagent, certain antibiotics and stains, etc. Such gloves should be worn when handling material containing category B pathogens. Suitable eye protection should also be made available on an individual basis to those persons wishing it. Suitable forms of respiratory protective devices should also be freely available.

Personal Hygiene. A high standard of personal hygiene must be maintained at all times. Hands must be washed after handling any infected material and always before leaving the laboratory. Hands must be washed after removing disposable gloves and protective clothing. Fingernails should be kept short and clean and brushed regularly when washing hands. To avoid the risk of contamination, long hair must be tied back or otherwise secured so it does not make contact with any work in progress. Any cuts and grazes on exposed parts of the body must be covered with waterproof dressings at all times. Adhesive labels must never be licked or other objects placed in the mouth; this applies to pipettes or pens, pencils, and other objects. Personal effects should not be brought into the laboratory, and pens, pencils, and markers used within the laboratory should always remain there. All nonessential paperwork and books should be kept away from those areas of the laboratory where potentially infective material is handled.

Food and Drink. Consumption of food or drink, smoking, and application of cosmetics are strictly forbidden in any laboratory.

Other Considerations

When tubes containing cultures are agitated, care should be taken to avoid contaminating the tube caps or cotton plugs. If the closure is contaminated, an aerosol can be generated when the cap or plug is removed.

Caution should be exercised in handling culture tubes and Petri dishes that are stored in incubators, refrigerators, and freezers. Broken tubes or dishes in any of these storage devices can result in the rapid dissemination of infectious aerosols. It is a good practice, if possible, to place glass culture containers in a secondary, unbreakable container.

If pathogenic microorganisms are going to be used in a school science laboratory, there must be some facility for sterilizing or disinfecting laboratory equipment. All material used must be sterilized even if it is going to be disposed of.

RULES FOR THE MICROBIOLOGICAL LABORATORY

Laboratory Safety Officer

At any premises where microbiological laboratory facilities exist, a safety officer responsible for microbiology should be designated as part of the overall safety organization. This safety officer must be sufficiently trained and experienced in microbiology to appreciate the risks of laboratory infection.

The safety officer must have direct access to the laboratory manager, scientist, or any other person with whom the major responsibility for laboratory safety lies; she should be informed when any new procedure or equipment is to be adopted.

The safety officer is responsible for keeping up-to-date in matters concerning microbiological safety. She must ensure that all new members of a staff are given adequate instruction before handling infectious material.

The safety officer must be informed promptly of any accidents or incidents, especially those involving infectious material, and she must ensure that written records of all such events are kept. She also must make sure that equipment such as safety cabinets and autoclaves is properly operated and maintained and that infectious material is rendered safe before disposal. The safety officer also sees to it that equipment to be repaired or serviced is disinfected if necessary before being handled by nonlaboratory personnel.

Model Rules for Students and Auxiliary Staff

PREVENTION OF INFECTION

1. Always wear the overall provided for your protection and see that it is properly fastened. Keep it apart from your outdoor clothing, not in your locker. Pegs should be provided. Do not take your overalls home to wash.
2. Do not wear your overall in the staff room or canteen. Take it off when you leave the laboratory.
3. Wash your hands often and always before leaving the laboratory or going to the staff room for food and drink or for a smoke. Cover cuts and scratches with waterproof dressings.
4. Do not eat, drink, smoke, or apply cosmetics in any laboratory. Use the staff room.
5. Do not touch any bottles, tubes, or dishes on any of the laboratory benches unless you have been told by the safety officer or your supervisor that it is safe for you to do so.
6. Do not dust or clean any work benches without permission from one of the laboratory staff.
7. If you have an accident of any kind or knock over or break any bottle, jar, tube, or other piece of equipment, tell the safety officer, your supervisor, or one of the laboratory staff at once.
8. Do not attempt to clean up after any accident without permission from a senior member of the laboratory staff. Do not pick up broken glass with your fingers. Use a dust pan and brush. Follow the instructions of senior staff members.
9. Do not enter any room with the red and yellow "Danger of Infection" sign on the door until the occupant tells you that it is safe to do so.
10. Do not empty any laboratory discard containers unless a label or an instruction says that you may do so.

If you work in the wash-up room, follow these instructions as well as those above:

1. Do not handle or wash any material that comes from the laboratory until it has been sterilized (autoclaved) or one of the laboratory staff or your supervisor has told you that it is safe to do so.
2. Do not place broken glass in plastic disposal bags. Use the labeled containers provided.
3. Do not work with the autoclave until you have been taught how to do so by your supervisor, and the safety officer is satisfied that you can competently operate it. Follow the written instructions displayed near it at all times.

If you cut or prick yourself or have any other accident that injures you, however slightly, report it to your supervisor at once and see that the safety officer records it in the accident book. This may save you a lot of trouble later.

If you obey these simple rules, you will be as safe as anyone else who works in the laboratory, but if you should fall ill, tell your doctor where you work so she can contact your supervisor if necessary. If in doubt about anything, ask the safety officer.

Supervision and Training

The person in charge of the laboratory supervises personnel and shares responsibility for their health and safety while at work. The employer should ensure that every person employed in a microbiology laboratory does read and understand the laboratory rules. The senior microbiologist must decide, in conjunction with senior staff in charge of microbiological laboratories, the techniques and procedures to be used. These should be distributed as a laboratory method handbook. As new hazards are recognized or equipment developed to conduct a procedure with greater safety, this handbook should be revised. When possible, techniques should be modified to avoid potentially dangerous manipulations and tests abandoned where risk, chemical, physical, or microbiological, cannot be justified.

Supervisory Staff

Supervisory staff should know and understand the details of all jobs laboratory personnel are expected to carry out and be aware of potential dangers. They should also ensure that staff members in their charge have received instructions on sound bacteriological techniques and the safe use of equipment. They must see to it that instructions are understood and followed.

Scientific Staff

All microbiology personnel must receive adequate instruction and training from a qualified microbiologist in the basic principles of good microbiological technique before they are allowed to handle contaminated material or cultures. Emphasis must be placed on the reasons behind good microbiological technique and the dangers inherent in any departure from it.

Auxiliary Staff

Training and supervision should extend to auxiliary staff (i.e., laboratory aides, maintenance and cleaning staff), who must be clearly informed of the extent of their duties and warned of the dangers of handling materials that they have not been specifically instructed to clean. A set of rules readily understood by all should be prepared; it should cover the duties undertaken by auxiliary staff, outlining their intent and implementation. A set of model rules is included earlier in this section.

Auxiliary staff should not be permitted to handle unsterilized, discarded material from any pathogen laboratory. They may, however, deal with discarded material from routine laboratories and operate equipment such as autoclaves after proper training.

As far as possible, the cleaning of any pathogen laboratory should be done by its staff. It may be desirable to allow the regular cleaning staff to clean floors and empty waste bins, but in category B pathogen laboratories this must be done under supervision. If supervision is not possible in general pathogen laboratories, then the following precautions must be observed:

1. No infectious material must be left accessible.
2. Material to be autoclaved must be placed where there is no danger of accidental spillage; it must be clearly identified.

Laboratory Access

A suitable, clearly defined, separate reception area should be provided that is located so as to obviate or minimize the need for access to the laboratory. It is recommended that whenever possible such areas not be inside pathogen laboratories.

Access to all areas of a microbiology laboratory by persons not employed in those sections should be discouraged. General pathogen and category B pathogen laboratories must be kept locked when unattended, with adequate arrangements for access in the event of an emergency. Untrained staff must not be allowed in the laboratory without competent supervision, which should continue until basic training has been completed.

Service agents and maintenance personnel, and any other individuals functioning at a similar level, must not be left in the laboratory without continuous supervision, as far as is reasonably practical, unless the laboratory has been made safe by the adequate containment of infectious materials, disinfection of surfaces, and clear and explicit instructions concerning the nature and extent of any work that such persons are permitted to carry out.

Health of Staff

Medical Fitness. Staff members must not have any medical condition that affects their ability to work in the laboratory environment. A medical adviser should be consulted before employing any person whose medical history or condition might conceivably give grounds for concern and in the case of any laboratory worker who develops a condition that places that person or fellow workers at risk.

Medical Monitoring. A record must be kept of the name, address, and telephone number of the personal physician of each staff member working in any pathogen laboratory. Staff in such laboratories should inform their doctors of the nature of their employment. Consider issuing a medical contact card stating that the bearer is employed in a microbiology laboratory.

When any staff member is out sick for more than three days without a satisfactory explanation, such absence must be reported to the department head. When laboratory infection is thought likely, the safety officer responsible for microbiology must see to it that the circumstances are investigated and the department head must make sure that the incident is reported to the person ultimately responsible for health and safety.

Preventive Inoculations. The advice of a medical adviser should be sought concerning preventive inoculations before any pathogenic organisms are handled.

Minor Injuries. Any injury sustained in the laboratory, however minor, must be recorded. Cuts and scratches on exposed parts of the body must be covered with waterproof dressings at all times.

Special Accommodation for Category B Organisms

1. A separate room must be provided for:
   (a) Handling, processing, or culturing potentially tuberculous material and mycobacteria.
   (b) Work, other than occasional routine isolation, with any of the organisms listed as category B in Table 7-2.
2. The same room may be used for any or all of these activities, but when it is used for work involving category B pathogens, work on other organisms should not be carried out at the same time. This minimizes the number of persons exposed to category B pathogens. When not in use for this purpose, it may serve as a general pathogen laboratory.
3. The room should be large enough to accommodate at least two persons, that is, not less than 800 ft^3. Ideally, the room should measure not less than 200 ft^2 (18 m^2).
4. The door(s) must be locked when the room is not in use. Doors must have glass panels so that the room's occupants can be seen from outside.
5. At least one exhaust protective cabinet (class 1) must be fitted.
6. Hand-washing facilities must be provided.
7. A warning sign, "Danger of Infection," with the international biohazard symbol must be displayed prominently on the outside of all doors accessing this room and on exhaust protective cabinets, incubators, refrigerators, and other equipment where category B microorganisms, agents, and materials are handled or stored.
8. A refrigerator and deep-freezer must be provided so that all category B specimens, materials, and reagents are stored in this room and nowhere else.
9. Access to this room must be restricted to persons authorized by the laboratory manager or the equivalent. Clerical or maintenance staff must not be permitted to enter until the environment has been made safe for them.

ANIMAL CARE AND FEEDING

All animals used in experimental research should be treated in a humane fashion. The animal care facilities should be kept clean, vermin controlled, wastes removed, and the concentration of pathogenic microorganisms kept to a minimum.

Animals can contract diseases from human beings. Such diseases as salmonellosis, influenza, tuberculosis, and infectious hepatitis can all be transmitted from man to animals. Anyone infected with a disease-causing microorganism must stay away from experimental animals.

Conversely, human beings can contract numerous diseases from infected animals. And infected animals can transmit disease to healthy animals. Any animal suspected of being ill should be isolated from other animals.

Animals should be handled very carefully. Students and instructors should never place their bare hands in an animal cage. Gloves should always be worn when handling animals. Cages should be sterilized before and after placing animals in them. All feeding devices and bedding materials must be sterilized as well. Access to animal care facilities must be strictly limited to those individuals directly responsible for their care.

Animal Quarters Employee Regulations

1. Wear protective clothing and use animal restraint equipment as instructed by your supervisor.
2. Report immediately all bites and scratches inflicted on you by an animal, and report all other injuries as promptly.
3. Keep your work area uncluttered. Allow sufficient aisle space between cage racks and work tables.
4. Do not smoke, eat, or drink in the animal quarters.
5. Set damaged or defective cages aside and label this equipment for repair.
6. Do not load carts too high, thus obscuring vision or exceeding the weight capacity of the cart.
7. Do not use the wrong transporter for the job. Cages that do not nest together when stacked in tiers will slide under impact and must be held in place by vertical or horizontal supports.
8. Broken glassware should be swept up with a brush and dustpan, not collected by hand.
9. Do not handle species that you have not been taught to handle. This is safer for you and the animal.

Personal Hygiene

1. Change into a clean uniform on arrival. Keep your street clothes in the locker room.
2. Keep your hands clean and your fingernails cut.
3. Shave daily before coming to work.
4. Do not take any personal items (combs, smoking tobacco, pipes, radios, chewing gum, coffee pots, etc.) into the animal rooms.
5. Wash your hands with soap and hot water under all of the following circumstances:
   (a) When leaving the locker room.
   (b) When leaving an animal room.
   (c) After handling dead or sick animals.
   (d) Immediately before starting work in an animal room.
6. Clothing soiled by animals should be removed and put in a laundry basket or autoclaved.

Animal Facilities

Categories for Animal Assignment

1. *Quarantine:* Elaborate control systems for isolating infected animals from the start. Involves a conditioning and treatment period; species housed separately; stringent precautions taken to control diseases.
2. *Research:* Animals may be allotted according to the following categories:
   (a) Acute (nonsurvival): hours or days; disease risk almost nil.
   (b) Acute (short-term survival): days or a few weeks; disease risk minimal.
   (c) Chronic (long-term survival): months or years; constant observation; maximum disease risk requires stringent control measures.
3. *Production:* Breeding of animals; time and space demands are great.

Floor Plans

1. *Single-corridor system:* All traffic uses one throughway; clean and dirty cages, healthy and sick animals, and food and personnel travel the same route, many times in direct contact; quarantine areas are difficult to provide for.
2. *Two-corridor system:* The flow of traffic is one way; when an entry into an animal room is made, the exit is via the dirty corridor; the standard sequence of animal cage routing would be (1) clean storage area, (2) clean corridor, (3) animal room, (4) dirty cages, dirty corridor, (5) dirty side of wash area, (6) clean side of wash area, (7) clean storage; air pressure dictates air flow from the clean corridor, to animal room, to dirty corridor.
3. *Three-corridor system:* A central corridor and two exterior dirty corridors or vice versa; basically the same as the two-corridor system, but with one-way floor patterns.

Structural Considerations

1. *Corridors:* Should be approximately 7 ft wide; floor wall junction should be covered to facilitate cleaning; guardrails or bumpers on equipment; access to plumbing and electricity.
2. *Animal room doors:* Should swing toward corridor; approximately 42 in. wide and 84 in. high, doors fit tightly; self-closing and equipped with kickplates and a viewing window.
3. *Floors:* Should be smooth, waterproof, nonabsorbent, nonslip, wear-resistant, acid- and solvent-resistant, capable of withstanding scrubbing with detergents and disinfectants; capable of supporting racks, equipment and storage areas without gouging, cracking, or pitting.
4. *Walls:* Should be smooth, waterproof, painted, joint-free, and capable of withstanding scrubbing with detergents and disinfectants.
5. *Ceilings:* Those formed by the concrete floor above are recommended if properly smoothed, sealed, and painted. Ceilings of plaster or firecode plasterboard should be sealed and painted with a washable finish; hung or false ceilings are unsatisfactory as are exposed pipes and fixtures.
6. *Fixtures:* Should be recessed and gasketed; electrical fixtures should be 4 ft from the floor and covered to protect them from water.
7. *Ventilation, temperature, and humidity control:* Air movement in one direction, filtered and conditioned, odorless. Temperature maintained at 72–80°F, with relative humidity of 35–70%; temperature and humidity should be controlled individually in each animal room or groups of rooms serving a common purpose.
8. *Lighting:* Should be uniformly diffused throughout the area; although 10- to 15-ft candles of light are sufficient to maintain vital ani-

mal activity and rhythms, at least 50-ft candles are necessary for the ordinary servicing of animal rooms. Animal treatment and examination areas should have a minimum of 100-ft candles at the work surface. There should be controlled, timed, on-off lighting systems.

9. *Drainage:* Drain pipes should be not less than 4 in. in diameter; for larger animals such as dogs in kennels, drains should be at least 6 in. in diameter. Floor drains are not essential in animal rooms for species such as rats, mice, or hamsters. The proper pitching of floors with drains is essential (0.25 in./ft). When not in use, drains should be capped and sealed.
10. *Storage areas:* The amount of space for food and bedding storage should be held to a minimum and arranged with constant turnover in mind. A separate area or room should be available for food, bedding, and equipment. Separate storage for animal waste and dead animals is essential.

Special Features

1. Automatic watering and waste-flushing systems
2. Cage and equipment washers
3. Incinerator
4. Autoclave
5. Quarantine room
6. Separate eating area
7. Vermin control
8. Routine employee medical examination and vaccinations
9. Protective equipment

8

COMPRESSED GASES AND CRYOGENIC MATERIALS

INTRODUCTION

A compressed gas as defined by the U.S. Department of Transportation (DOT), which regulates the commercial transport of such gases, is any material or mixture having in the container either an absolute pressure greater than 40 psi (40 lbf/in.^{2} or 276 kPa) at 21°C, or an absolute pressure greater than 104 psi (717 kPa) at 54°C, or both, or any liquid flammable material having a Reid vapor pressure greater than 40 psi (276 kPa) at 38°C.

Compressed gases present several kinds of hazards:

Fire

Toxicity

Corrosiveness

Energy release due to pressure

CYLINDERS OF COMPRESSED GASES

Gaseous reagents are usually stored as a compressed gas in heavy metal cylinders of various sizes. The gas is released through a high-pressure valve that is protected by a domed cover when the cylinder is not in use. The cylinders are color-coded and labeled. With very few exceptions, gases are supplied at very high pressures (1000–2000 psi).

DOT has established codes that specify the materials of construction and the capacities, test procedures, and service pressures of the cylinders in which compressed gases are stored. A common compressed-gas steel cylinder might be designated DOT 3A-2000, which indicates that it has been manufactured under specification 3A and has an operating pressure of 2000 psi (13.8 MPa) at 21°C. The typical weight of a size-2 cylinder of gas is 45.5 kg (100 lbs). The tank is about 48 in. high and 10 in. in diameter (122 x 25.4 cm).

However, regardless of the pressure rating of a cylinder, the physical state of the material within it determines the pressure of the gas. For example, liquefied gases such as propane or ammonia will exert their own vapor pressure as long as any liquid remains in the cylinder and the critical temperature is not exceeded.

Proper procedures for the use of compressed-gas cylinders in the laboratory include attention to proper identification, handling and use, transportation, storage, and the return of the empty cylinder.

Identification of Contents

The contents of any compressed-gas cylinder should be clearly identified so they can be recognized easily, quickly, and completely by any laboratory worker. Such identification should be stenciled or stamped on the cylinder itself or a label should be provided that cannot be removed from the cylinder (three-part tag systems, which are available commercially, can be very useful for identification and inventory). No compressed-gas cylinder should be accepted for use that does not legibly identify its contents by name. Color coding is helpful but by itself is not a reliable means of identification; cylinder colors vary from supplier to supplier and labels on caps have no value as caps are interchangeable. If the labeling on a cylinder becomes unclear or an attached tag is defaced, so that the contents cannot be identified, the cylinder should be marked "contents unknown" and returned directly to the manufacturer.

All gas lines leading from a compressed-gas supply should be clearly labeled so as to identify the gas, the laboratory served, and relevant emergency telephone numbers. The labels should be color-coded to distinguish hazardous gases (such as flammable, toxic, or corrosive substances) with, for example, a yellow background and black letters from safe (inert) gases with, for example, a green background and black letters. In areas where flammable compressed gases are stored, signs should be conspicuously posted identifying the substances and appropriate precautions (e.g., HYDROGEN-FLAMMABLE GAS, NO SMOKING, NO OPEN FLAMES).

Non-Vendor-Owned Pressure Vessels

Each pressure vessel should have stamped on it (or on an attached plate) its basic allowable working pressure, the allowable temperature at this pressure, and the material of construction. Similarly, the relieving pressure and setting data should be stamped on a metal tag attached to installed pressure-relieving devices, and the setting mechanisms should be sealed. Relief devices used on pressure regulators do not require these seals or numbers.

HANDLING AND USE OF COMPRESSED GAS TANKS

General Rules for Handling Gas Cylinders

1. Compressed-gas cylinders not currently being used or those purchased in excess of immediate needs should be stored in a storage shed. This is especially desirable for such toxic or flammable gases as ammonia, carbon monoxide, chlorine, hydrogen, hydrogen chloride, hydrogen cyanide, hydrogen sulfide, phosgene, etc.
2. For gas cylinders in use, adhere to the following safety rules:
   (a) Make sure the cylinders are firmly secure; use commercially available gas cylinder supports or have a rack made. The following incident points out the potential dangers involved: A small carbon dioxide cylinder fell from a dock, causing the valve to break off. The cylinder took off like a rocket, traveled 115 ft, struck a piling and was deflected, traveled another 90 ft, reaching a height of 30 ft.
   (b) Do not rely solely on a color code for identification of a gas. Instead, be guided by its tag or decal.
   (c) Mark each regulator with the symbol of the gas with which it is to be used. Don't interchange them. Seek advice from your supervisor before using adapters. Even special "corrosion-resistant" regulators for halogen gases *do* corrode and clog when kept on the tank for prolonged periods. These regulators should be removed after each use and the caps replaced on the tanks. Monel needle valves should be removed, rinsed with water, and dried with ethanol after each use.
   (d) Keep a crescent wrench immediately available to close cylinders not equipped with valve handles.
   (e) *When contents are toxic, order small-size cylinders only.* Do not store them in the hallway; secure them firmly and use them only in a hood.
   (f) When a flow gauge is used in place of the low-pressure gauge of a regulator in a closed system, a pressure release should be incorporated.
   (g) Transport cylinders on carts—*keep caps in place.*
   (h) Stand away from the face of the regulator when opening the valve. Free gases should be turned on slowly and fully. Liquid gases should be turned on partially.
   (i) Never "crack" a gas cylinder to determine whether the cylinder is empty or full without using a pressure regulator and gauge.
   (j) On oxygen cylinders, ordinarily innocuous oils and greases may become combustible and should not be used on pressure regulators, gauges, or other fittings.

Transportation

Cylinders should be strapped and transported on two-wheeled carts. At their destination, they should be securely attached to the wall with straps or chains. Devices are available for attaching tanks to lab benchtops.

Securing

Compressed-gas cylinders should be firmly secured at all times. A clamp and belt or chain are generally suitable for this purpose. Pressure-relief devices protecting equipment that is attached to cylinders of flammable, toxic, or otherwise hazardous gases should be vented to a safe place.

Shutdown

High-pressure systems should be shut down when not in use. To properly shut down such a system, perform the following:

1. Close the high-pressure valve (clockwise for tightness).
2. Open the needle valve on the delivery side to bleed gas out of the regulator (both gauges should read zero).

3. Turn the regulator valve counterclockwise until it is loose.
4. Close the needle delivery valve.

Empty Cylinders

A cylinder should never be emptied to a pressure lower than 25 psi (172 kPa) because the residual contents may become contaminated if the valve is left open. Empty cylinders should not be refilled. Rather, the regulator should be removed and the valve cap replaced. The cylinder should be clearly marked as "empty" and returned to a storage area for pickup by the supplier. Empty and full cylinders should not be stored in the same place.

Cylinder discharge lines should be equipped with approved check valves to prevent the inadvertent contamination of cylinders that are connected to a closed system where the possibility of flow reversal exists. Sucking back is particularly troublesome in the case of gases used as reactants in a closed system. A cylinder in such a system should be shut off and removed from the system when the pressure remaining in the cylinder is at least 25 psi. If there is a possibility that a cylinder has been contaminated, the cylinder in question should be so labeled and returned to the supplier.

Acetylene tanks must always be used in an upright position. Acetylene is dissolved in an organic solvent (hexane) in the tank to reduce the explosion hazard. If the cylinder is used in anything but an upright position, the hexane may be expelled.

Compressed-Gas Cylinder Venting

Occasionally, it is necessary to vent leaking or damaged cylinders of compressed gases. Such gases may be toxic, flammable, or corrosive. This procedure should be carried out by personnel trained in the safe handling of compressed-gas cylinders. Safety equipment such as respirators should be available, and adequate ventilation to avoid exposure or explosion provided. Disposal facilities equipped with scrubbing or incinerating devices may be required, depending on such factors as the volume and nature of the material, the physical location, and local regulations. Venting should be done at a rate that will not cause environmental and safety problems and in compliance with local emission regulations.

Leaking Compressed-Gas Cylinders

Occasionally, a cylinder or one of its component parts develops a leak. Most of these leaks occur at the top of cylinder in areas such as the valve threads, safety devices, valve stem, and valve outlet.

If a leak is suspected, do not use a flame for detection; rather, use a flammable-gas leak detector, soapy water, or an alternatively appropriate solution. If the leak cannot be remedied by tightening a valve gland or packing nut, emergency action procedures should be instituted and the supplier notified. Laboratory workers should never attempt to repair a leak in valve threads or a safety device; rather, they should consult with the supplier for instructions.

The following general procedures can be used for leaks of minimum size when the indicated action can be taken without risking the serious exposure of personnel. If it is necessary to move a leaking cylinder through populated portions of the building, place a plastic bag, rubber shroud, or similar device over the top and tape it (duct tape preferred) to the cylinder to confine the leaking gas.

1. In the case of flammable, inert, or oxidizing gases, move the cylinder to an isolated area (away from combustible material if the gas is flammable or an oxidizing agent) and post signs that describe the hazards and state warnings.
2. Corrosive gases may increase the size of the leak as they are released and some corrosives are also oxidants or flammable. Move the cylinder to an isolated, well-ventilated area and use appropriate means to direct the gas into a suitable chemical neutralizer. Post signs that describe the hazards and state warnings.
3. For toxic gases, follow the same procedure as for corrosive gases. Move the cylinder to an isolated, well-ventilated area and use appropriate means to direct the gas into a suitable chemical neutralizer. Post signs that describe the hazards and state the warnings.

When the nature of the leaking gas or the size of the leak constitutes a more serious hazard, self-contained breathing apparatus or protective apparel, or both, may be required. Basic action for large or uncontrolled leaks may include any of the following steps:

1. Evacuation of personnel
2. Rescue of injured personnel by crews equipped with adequate personal protective apparel and breathing apparatus
3. Firefighting action
4. Emergency repair
5. Decontamination

ANCILLARY EQUIPMENT

Valves

Standard cylinder-valve outlet connections have been devised by the Compressed Gas Association (CGA) to prevent the mixing of incompatible gases (see such a list in *Academic Laboratory Chemical Hazards Guidebook*, Chap. 1) due to an interchange of connections. The outlet threads used vary in diameter; some are internal and some are external; some are right-handed and some are left-handed. In general, right-handed threads are used for nonfuel and water-pumped gases, and left-handed threads for fuel and oil-pumped gases. Information on the standard equipment assemblies used with specific compressed gases is available from the supplier. To minimize undesirable connections that may result in a hazard, only CGA standard combinations of valves and fittings should be used in compressed-gas installations; the assembly of miscellaneous parts (even of standard approved types) should be avoided. The threads on cylinder valves, regulators, and other fittings should be examined to ensure that they correspond to one another and are undamaged.

Cylinders should be placed so that the cylinder valve is accessible at all times. The main cylinder valve should be closed as soon as it is no longer necessary that it be open (i.e., it should never be left open when equipment is unattended or not operating). This precaution is necessary not only for safety reasons when the cylinder is under pressure, but also to prevent the corrosion and contamination that could result from the diffusion of air and moisture into the cylinder after it has been emptied.

Most cylinders are equipped with hand-wheel valves. Those that are not should have a spindle key on the valve spindle or stem while the cylinder is in service. Only wrenches or other tools provided by the cylinder supplier should be used to open a valve. In no case, should pliers be used to open a cylinder valve. Some valves require washers, and this should be checked before the regulator is fitted.

Cylinder valves should be opened slowly; the valve on an unregulated cylinder should never be "cracked." At no time is it necessary to open the main cylinder valve all the way; the resulting flow will be much greater than one would ever want. It is safe practice to open the main valve only to the extent necessary.

When opening the valve on a cylinder containing an irritating or toxic gas, the user should try to stand away from the fittings most likely to leak (usually, the highest-pressure fittings). If there is appreciable air flow around the cylinder, stand on the upwind side of the cylinder with the valve pointed downwind. Warn those working nearby in case of a possible leak.

Relief Valves

Compressed-gas cylinders are designed to operate safely up to 51.7°C (125°F). When they become hot, the pressure rises and a safety valve releases the contents. Otherwise, the tank would explode like a bomb. If a fire breaks out in a system connected to a tank, shut off the tank valve before attempting to extinguish the fire.

Pressure Regulators

Pressure regulators are generally used to reduce a high-pressure supplied gas to a desirable lower pressure and to maintain a satisfactory delivery pressure and flow level for the required operating conditions. They can be obtained to fit many requirements, including a range of supply and delivery pressures, flow capacity, and construction materials. All regulators are of the diaphragm type, spring-loaded or gas-loaded, depending on pressure requirements. They may be single-stage or two-stage.

High-pressure tanks are ordinarily used with a two-stage regulator with appropriate pressure ranges. Each regulator is supplied with a specific CGA standard inlet connection to fit the outlet connection on the cylinder valve for that particular gas. Attachment fittings on regulators must match the tank valve. Adapters are available, but their use should be discouraged. The same regulator should not be used with different gases. *Never* use a regulator that has been used for some other gas on an oxygen cylinder. High-pressure oxygen can react explosively with oil residues from other gases that have been transferred to the regulator. Regulators are attached to the valve of the tank, with a thin piece of plumber's teflon tape as a lubricant and seal. As a general rule, oil should not be used on regulator valves or cylinder valves. For oxygen tanks, the regulator threads must be free of all organic materials.

All pressure regulators should be equipped with spring-loaded pressure-relief valves to protect the low-pressure side. When used on cylinders of flammable, toxic, or otherwise hazardous gases, the relief valve should be vented to a safe location. The use of internal-bleed-type regulators should be avoided.

Regulators for use with noncorrosive gases are usually made of brass. Special regulators made of corrosion-resistant materials can be obtained for use with such gases as ammonia, boron trifluoride, chlorine, hydrogen chloride, hydrogen sulfide, and sulfur dioxide. Because of freeze-up and corrosion problems, regulators used with carbon dioxide gas must have special internal design features and be made of special construction materials.

When the regulator is attached and before opening the tank valve, be sure the valve for regulating the outlet pressure is turned counterclockwise until there is no resistance. Then open the tank valve by turning it counterclockwise. Do *not* stand directly in front of a regulator during this step. An explosion or fire can blow out the face of the regulator gauges. If anything is to go wrong, it will most likely happen at this point. Give the tank valve a full counterclockwise turn and then reverse it one-half turn clockwise. When the tank regulator shows the tank pressure, check for leaks in the valves with a soap solution and tighten them if necessary. Open the valve of the low-pressure stage of the regulator (clockwise) to deliver gas at the indicated pressure. Open the needle valve to deliver gas to the system as desired.

Pressure Gauges

The proper choice and use of a pressure gauge are the responsibility of the user. Among the factors to be considered are the flammability, compressibility, corrosiveness, toxicity, temperature, and pressure range of the fluid with which it is to be used.

A pressure gauge is normally a weak point in any pressure system because its measuring element must operate in the elastic zone of the metal involved. The resulting limited factor of safety makes careful gauge selection and use mandatory and often dictates the use of accessory protective equipment. The primary element of the most commonly used gauges is a Bourdon tube, which is usually made of brass or bronze and has soft-soldered connections. More expensive gauges can be obtained that have Bourdon tubes made of steel, stainless steel, or other special metals and welded or silver-soldered connections.

Consideration should be given to alternative methods of pressure measurement that may provide greater safety than the direct use of pressure gauges. Such methods include the use of seals or other isolating devices in pressure tap lines, indirect-observation devices, and remote measurement by strain-gauge transducers.

FLAMMABLE GASES

Sparks and flames should be kept from the vicinity of cylinders of flammable gases. An open flame should never be used to detect leaks of flammable gases. Rather, soap water should be used, except during freezing weather, when a 50% glycerine-water solution or its equivalent may be used. Connections to piping, regulators, and other appliances should always be kept tight to prevent leakage, and the hoses used should be maintained in good condition. Regulators, hoses, and other appliances used with cylinders of flammable gases should not be interchanged with similar equipment intended for use with other gases.

All cylinders containing flammable gases should be stored in a well-ventilated place. Reserve stocks of such cylinders should never be stored in the vicinity of cylinders containing oxygen.

LIQUEFIED GASES AND CRYOGENIC LIQUIDS

The primary hazards of cryogenic liquids are fire or explosion, pressure buildup, embrittlement of structural materials, contact with and destruction of living tissue, and asphyxiation. The fire or explosion hazard is obvious when gases such as hydrogen, methane, and acetylene are used. Enriched oxygen will greatly increase the flammability of ordinary combustible materials and may even cause some noncombustible materials (such as carbon steel) to burn readily.

Oxygen-saturated wood and asphalt have been known to literally explode when subjected to shock. Because it has a higher boiling point (−183°C) than nitrogen (−195°C), helium (−269°C), or hydrogen (−252.7°C), oxygen can be condensed out of the atmosphere during the use of these lower-boiling-point cryogenic liquids. Particularly with liquid hydrogen, conditions may exist for an explosion.

Even very brief skin contact with a cryogenic liquid is capable of causing tissue damage similar to that of thermal burns, and prolonged contact may result in blood clots with potentially very serious consequences.

Eye protection, preferably a face shield, should be worn when handling liquefied gases and other cryogenic fluids. Gloves should be chosen that are impervious to the fluid being handled and loose enough to be removed quickly. A potholder may be a desirable alternative. The area should be well ventilated. The transfer of liquefied gases from one container to another should not be attempted for the first time without the direct supervision and instruction of someone experienced in this operation.

It is advisable to furnish all cylinders and equipment used to contain liquefied gases that are not vendor-owned with a spring-loaded pressure-relief device (not a rupture disk) because of the magnitude of the potential hazard, the large amount of flammable or toxic (or both) gas that can result from activation of a nonresetting relief device on a container of liquefied gas. Commercial cylinders of liquefied gases are normally supplied with only a fusible-plug type of relief device, as permitted by DOT regulations.

Cylinders and other pressure vessels used for the storage and handling of liquefied gases should not be filled to more than 80% capacity. This is a precaution against the possible thermal expansion of the contents and bursting of the vessel by hydrostatic pressure. If the possibility exists that the temperature of the full cylinder might be increased to above 30°C, a lower percentage (e.g., 60%) of capacity should be established as the limit.

LOW-TEMPERATURE EQUIPMENT

At low temperatures, the impact strength of ordinary carbon steel is greatly reduced. The steel may fail when subjected to impact or mechanical shock, even though its ability to withstand slowly applied loading is not impaired. This type of failure normally occurs at points of high stress (such as at notches in the material or abrupt changes of section).

The 18% chromium–8% nickel stainless steels retain their impact resistance down to approximately −240°C, the exact value depending heavily on special design considerations. The impact resistance of aluminum, copper, nickel, and many other nonferrous metals and alloys increases with decreasing temperatures.

COLD TRAPS

Cold traps should be of sufficient size and low enough temperature to collect all condensable vapors present in a vacuum system and should be interposed between the system and the vacuum pump. They should be checked frequently to guard against their becoming plugged by the freezing of material collected in them.

The common practice of using acetone-dry ice as a coolant should be avoided. Isopropanol or ethanol work as well as acetone and are cheaper, less toxic, less flammable, and less prone to foam on the addition of small particles of dry ice. Dry ice and liquefied gases used in refrigerant baths should always be open to the atmosphere; they should never be used in closed systems where they could develop uncontrolled and dangerously high pressures.

After the completion of an operation in which a cold trap has been used, the system should be vented. This venting is important because volatile substances that have collected in the trap may vaporize when the coolant has evaporated and cause a pressure buildup that could blow the apparatus apart. In addition, the oil from some pumps can be sucked back into the system.

Extreme caution should be exercised in using liquid nitrogen as a coolant for a cold trap. If such a system is opened while the cooling bath is still in contact with the trap, oxygen may condense from the atmosphere, which, if the trap contains organic material, will create a highly explosive mixture. Thus, a system connected to a liquid nitrogen trap should not be opened to the atmosphere until the trap has been removed. Also, if the system is closed after even a brief exposure to the atmosphere, some oxygen (or argon) may have already condensed. Then, when the liquid nitrogen bath is removed or when it evaporates, the condensed gases will vaporize with attendant pressure buildup and potential blowup.

PRESSURIZED SYSTEMS

Pressure System Design Requirements

The philosophy underlying pressure system design is quite simple. A pressurized operation should take place in manned areas only when it is demonstrated to be safe. If safety cannot be demonstrated, unmanned operation is mandatory. Even then, the consequences of damage to facilities and equipment must be carefully assessed. Therefore, the goals of design are twofold, in order of priority: (1) to operate safely and (2) to attain intended system objectives. The most essential ingredients in achieving safety in pressure systems are a knowledge of the potential hazards and a healthy respect for them. The significant points are as follows:

1. Minimize risk and exposure. The goal of minimum risk and exposure is a self-evident and worthwhile principle. There are many ways to minimize risks and exposure; only a few will be listed and briefly discussed.
2. Identify all hazards and consequences.The nature and magnitude of potential hazards may have a significant bearing on decisions, so the early, methodical identification of all hazards is a necessity. In particular, consideration must be given to how failures take place and their consequences.
3. Go "remote." The location of the pressure operation has much to do with determining the degree of hazard. So also do the frequency and duration of personnel presence. The best possible locations are those remote from all personnel.

4. Minimize the pressure and volume. Stored energy available for release in case of the sudden rupture of a pressure vessel is proportional to total volume and pressure. Unnecessary risks are created by the use of a volume or pressure greater than required.
5. Abide by recognized standards. Standards are available for the design of a system. These derive from analytical procedures and experience. Observance of existing standards provides the least costly assurance that a system is safe.
   (a) ASME Boiler and Pressure Vessel Code Division I, Section VII.
   (b) DOT Tariff 32 or current version.
   (c) American National Standards Institute (ANSI) (ANSI standards are of particular use in piping design).
   (d) California supplements to the ASME code.
   (e) Fire Underwriter rules.
6. Design conservatively. When in doubt, be conservative in judgment. Do not lean too heavily on what is thought to be an "inherent safety factor." Sometimes, this supposed factor is not present. The MAWP (maximum allowable working pressure) should be about 15% above the desired maximum working pressure. The MAWP is then used to specify the design strength of the vessel. Relief valves should relieve any pressures above MAWP. Pressure-proof tests should be conducted at 1.5 times the MAWP. A factor of safety four times the MAWP, based on rupture, is established as a minimum for all pressure vessels and other pressure components to be used in manned areas.
7. Use material with a predictably safe failure mode. This injunction is aimed primarily at "brittle" materials, which sometimes totally fail in an unpredictable fashion. Seek materials that will fail in a ductile manner. A brittle material should never be used in a manned area unless it is properly shielded or barricaded. This selection is, if anything, more important than the choice of a suitable factor of safety. However conservative one's intent, a failure is possible if the material was not correct for the application, fabrication was haphazard, or the right processes were not employed (i.e., heat treatment or stress relief).
8. Demonstrate structural integrity by a proof test. Pressure vessels should be placed into service only after passing appropriate proof tests. These tests are conducted at a level exceeding the MAWP. Sometimes, it is required that this test be conducted at an operating temperature extreme because of the temperature dependence of material properties.
9. Operate within the original design intent. Do not exceed the MAWP. Also, do not change working fluids or service environments without taking possible harmful effects into consideration.
10. Provide backup protection. Suitable pressure relief devices should be installed at appropriate locations in all pressure systems to assure that the pressure level will stay within predetermined safe limits in spite of possible equipment malfunction or operational errors. Redundancy in relief devices is recommended for the more hazardous systems.
11. Use quality hardware. Quality hardware is an obvious necessity. Obtaining quality hardware often turns out to be a most difficult requirement with which to comply. Insist on information about the manufacturer's design, test, and fabrication of hardware and its quality-control procedures.
12. Use protective shields. The use of protective shields is a practical way around many problem areas. Although protective shields are not a total solution, the need for them is quite apparent in circumstances where (unavoidably) material has an unpredictable failure mode, where the quality of critical hardware is unknown, and where not all conventional criteria have been met (e.g., an uncertain factor of safety).
13. Use tie downs. The tie down of tubes, hoses, and piping at frequent intervals can prevent serious injury. (Should a line fail under pressure, it can "whip" unless restrained.)

Assembly and Operation

During the assembly of pressure equipment and piping, only appropriate components should be used and care taken to avoid strains and concealed fractures resulting from the use of improper tools or excessive force. Piping should not be used to support equipment of any significant weight.

All-brass and stainless steel fittings should be used with copper or brass and steel or stainless steel tubing, respectively. It is very important that fittings of this type be installed correctly. It is not usually advisable to mix different types of fittings in the same apparatus assembly.

In assembling copper-tubing installations, sharp bends should be avoided and considerable flexibility should be allowed. Copper-tubing work hardens and cracks on repeated bending. It should be inspected frequently and replaced when necessary.

Threads that do not fit accurately should not be forced. Thread connections should match correctly; tapered pipe threads cannot be joined with parallel machine threads. A suitable thread lubricant should be used when assembling the apparatus. However, an oil or lubricant must never be used on any equipment that will be used with oxygen. Parts with damaged or partly stripped threads should be rejected.

Stuffing boxes and gland joints are a likely source of trouble in pressure installations. Particular attention should be given to the proper installation and maintenance of these parts, including the proper choice of a lubricant and packing material.

Experiments carried out in closed systems and involving highly reactive materials, such as those subject to rapid polymerization (e.g., dienes or unsaturated aldehydes, ketones, or alcohols), should be preceded by small-scale tests using the exact reaction materials to determine the possibility of an unexpectedly rapid reaction or unforeseen side reactions. All reactions under pressure should be shielded.

Autoclaves and other pressure-reaction vessels should not be filled more than half full to ensure that space remains for the expansion of the liquid when it is heated. Leak corrections or adjustments to the apparatus should not be attempted while it is pressurized; rather, the system should be depressurized before mechanical adjustments are made.

Immediately after an experiment with low-pressure equipment connected to a source of high pressure is concluded, the low-pressure equipment should either be disconnected entirely or left independently vented to the atmosphere. This will prevent the gradual buildup of excessive pressure in the low-pressure equipment due to leakage from the high-pressure side.

Vessels or equipment made partly or entirely of silver, copper, or alloys containing more than 50% copper should not be used in contact with acetylene or ammonia. Those made of metals susceptible to amalgamation (such as copper, brass, zinc, tin, silver, lead, and gold) should not come into contact with mercury. This includes equipment that has soldered and brazed joints.

Prominent warning signs should be placed in any area where a pressure reaction is in progress so that others entering the location will be aware of the potential hazard.

Testing Pressure Equipment

All pressure equipment should be tested or inspected periodically. The interval between tests or inspections is determined by the severity of the service involved. Corrosive or otherwise hazardous service will require more frequent tests and inspections. Inspection data should be stamped on or attached to the equipment.

The use of a soap solution and air or nitrogen pressure to the maximum allowable working pressure of the weakest section of the assembled apparatus is usually an adequate means of testing a system for leaks through threaded joints, packings, valves, and such.

Before any pressure equipment is altered, repaired, stored, or shipped, it should be vented and all toxic or other hazardous material removed completely so it can be handled safely. Especially hazardous materials may require special cleaning techniques.

Hydrogen Embrittlement

Hydrogen is a powerful promoter of steel embrittlement, due to its ability to diffuse through the iron lattice. Materials that are embrittled show an erratic drop in tensile strength, have difficulty in deforming or adjusting to residual stresses, and will be more susceptible to cracking. It is possible for a steel to display normal properties in a number of regularly conducted mechanical tests, and yet fail by hydrogen embrittlement while sustaining a static tensile load of less then yield-point value. This is called delayed brittle failure.

Special alloy steels should be used for liquids or gases containing hydrogen at temperatures greater than 200°C or at pressures greater than 500 psi (34.5 MPa) because of the danger of weakening carbon steel equipment by hydrogen embrittlement. In general, chromium-molybdenum steels should be used in hydrogen atmospheres at elevated temperatures and pressures. 304 stainless steel alloy is suitable up to 800°F. Some other suitable alloys are 347 Alt and A-286.

Pressure-Relief Devices

All pressure or vacuum systems and all vessels that may be subjected to pressure or vacuum should be protected by pressure-relief devices. Experiments involving highly reactive materials that might explode may also require the use of special pressure-relief devices, operating at a fraction of the permissible working pressure of the system.

Examples of pressure-relief devices include the rupture-disk type used with closed-system vessels and the spring-loaded relief valves used with vessels for transferring liquefied gases. The following considerations are advisable in the use of pressure-relief devices:

1. The maximum setting of a pressure-relief device is the rated working pressure established for the vessel or for the weakest member of the pressure system at the operating temperature. The operating pressure should be less than the allowable working pressure of the system. In the case of a system protected by a spring-loaded relief device, the maximum operating pressure should be from 5 to 25% lower than the rated working pressure, depending on the type of safety valve and the importance of leak-free operation. In the case of a system protected by a rupture-disk device, the maximum operating pressure should be about two-thirds that of the rated working pressure; the exact figure is governed by the fatigue life of the disk used, the temperature, and load pulsations.
2. Pressure-relief devices that may discharge toxic, corrosive, flammable, or otherwise hazardous or noxious materials should be vented in a safe and environmentally acceptable manner.
3. Shutoff valves should not be installed between pressure-relief devices and the equipment they are to protect.
4. Only qualified persons should perform maintenance work on pressure-relief devices.
5. Pressure-relief devices should be inspected periodically.

Glass Equipment

The use of glassware for work at pressure extremes should be avoided whenever possible. Glass is a brittle material subject to unexpected failures due to factors such as mechanical impact and assembly and tightening stresses. Glass equipment, such as rotameters and liquid-level gauges, that is incorporated in metallic pressure systems should be installed with shutoff valves at both ends to control the discharge of liquid or gaseous materials in the event of breakage.

Glass equipment in pressure or vacuum service should be provided with adequate shielding to protect users and others in the area from flying glass and the contents of the equipment. New or repaired glass equipment for pressure or vacuum work should be examined for flaws and strains under polarized light.

Corks, rubber stoppers, and rubber or plastic tubing should not be relied on as relief devices for the protection of glassware against excess pressure; a liquid seal, Bunsen tube, or equivalent positive relief device should be used. When glass pipe is employed, only proper metal fittings should be used.

Plastic Equipment

Except as noted below, the use of plastic equipment for pressure or vacuum work should be avoided unless no suitable substitute is available.

Tygon and similar plastic tubing have some limited applications in pressure work. These materials can be used for natural gas, hydrocarbons, and most aqueous solutions at room temperature and moderate pressure. The details of permissible operating conditions must be obtained from the manufacturer. Because of their very large coefficients of thermal expansion, some polymers have a tendency to expand a great deal on heating. Thus, if the valve or joint is tightened when the apparatus is cold, the plastic can entirely close an opening when the temperature increases. This problem can be a hazard in equipment subjected to very low temperatures or to alternating low and high temperatures.

Pressure System Checklists

Design

1. Determine operating conditions:
   (a) Operating pressure.
   (b) Normal operating temperatures and temperature extremes.
   (c) Cyclic nature of pressure loading.
   (d) Presence of hostile environments.
2. Establish the MAWP.
3. Identify potential hazards.
4. Minimize pressurized volume.
5. Design to appropriate standards. Preferred designs are based on Division 1, Section VIII, ASME Boiler and Pressure Vessel Code. Other standards also exist.
6. Materials and processes. Materials must:
   (a) Have predictable failure modes.
   (b) Be chemically compatible.
7. Establish and design to a minimum factor of

safety equal to 4, higher if hazardous conditions warrant it.

8. Document all calculations.

Fabrication

1. Material certifications.
2. Proof-tested. The procedure should be completely specified:
   (a) Test level (usually, but not always, to 1.5 times the MAWP).
   (b) Holding time, maximum pressure.
   (c) Temperature during testing.
   (d) Additional measurements and methodology if desired.
   (e) Whether a representative need be present to witness the test.
   (f) Any special requirements.
3. Specify all paperwork, including supplementary notes and certifications that may be required.

Operations

1. Train all personnel in operating and emergency procedures.
2. Verify the system: all components calibrated and certified.
3. Resolve all toxicity and/or flammability problems.
4. Approved the standard operating procedure, if required.
5. Emergency shutdown procedures should be written down.
6. Relief devices tagged and in the recall system.
7. Complete package of system information on file (including proof-test data).
8. System registered with safety officer or committee.
9. Service log, if required.

SUGGESTED READINGS

Compressed Gas Association, *Characteristics and Safe Handling of Medical Gases,* New York, 1965, Pamphlet P2.

Compressed Gas Association, *Safe Handling of Compressed Gases in Containers,* New York, 1965, Pamphlet P1.

Handbook of Compressed Gases, Van Nostrand Reinhold, New York, 1981.

Loomis, A.W. (ed.), *Compressed Air and Gas Data,* Ingersoll-Rand, Washington, N.J., 1980.

9

VENTILATION

INTRODUCTION

Laboratories that use chemicals vary from spacious well-designed facilities to those that consist of a single room designated as a laboratory, with little or no provision for ventilation. The need for ventilation of these different types of laboratories will vary from the provision of simple comfort for the occupants to the control of highly toxic volatile substances. This chapter will center on ventilation for the control of toxic chemicals; however, the overall performance of laboratory workers will also benefit from ventilation systems that control the temperature, humidity, and concentration of odoriferous materials in the laboratory.

There is a tendency for laboratory workers to associate odor with toxicity. This tendency may result in overconcern for an odoriferous substance of low toxicity and a lack of concern for highly toxic substances that have little or no odor.

The steady increase in the cost of energy in recent years has resulted in a conflict between the desire to minimize the costs of heating or cooling and dehumidifying laboratory air and the need to provide laboratory workers with improved ventilation as a means of protection from toxic gases, vapors, aerosols, and dusts. Although the energy costs associated with tempering the input air for laboratories are often substantial, cost considerations should never take precedence over ensuring that laboratories have adequate ventilation systems to protect workers from hazardous concentrations of airborne toxic substances. Thus, any changes in overall laboratory ventilation systems to conserve energy should be instituted only after the thorough testing of their effects has demonstrated that laboratory workers will still have adequate protection. An inadequate ventilation system can be worse than none, because it is likely to give laboratory workers an unwarranted sense of security that they are protected from airborne toxic substances.

GENERAL CONCEPTS

Heating, ventilation, and air-conditioning systems are integral parts of school building design and are often much more suitable for ordinary classrooms than for laboratories. In older schools, steam radiators and window sashes constitute a simple arrangement. In newer buildings, complex air-handling systems are commonly utilized.

In recent years, energy conservation objectives favor recirculation of building air with minimal exhaust or fresh air intake. A central ventilation system may distribute throughout the building any undesirable gas or vapor spilled or released in the laboratory area unless effective isolation and local ventilation are provided.

Local exhaust ventilation taking suction as near as possible to the source of release constitutes the most energy-efficient as well as most effective means of control. Laboratory fume hoods and ventilated cabinets are commonly utilized in laboratories to achieve local removal. Conversely, the task of dilution or replacement of contaminated air by sweeping with large amounts of fresh air may be difficult, slow, expensive, and energy wasteful.

The discussion here will primarily address the means of achieving effective local ventilation within the laboratory. The objectives of such ventilation systems are:

1. To remove toxic gas, vapor, or dust and thus reduce the possibility of harmful health effects.
2. To remove flammable vapor or gas and thus reduce the possibility of fire or explosion.
3. To remove undesirable odor even though it may be relatively harmless.
4. To remove heat and humidity generated in the laboratory or shop.

Threshold Limit Value

Through experience and animal testing, reasonably safe concentrations in air or threshold limit values (TLV) have been established and published by the American Conference of Governmental Industrial Hygienists (ACGIH). These time-weighted average concentrations are expressed either as parts per million parts of air by volume (ppm) or as milligrams per cubic meter (mg/m^3). They are not precise, but are used as guides in evaluating or controlling the exposure of workers to contaminants in plants, including laboratories. Thus, a low TLV would mean that the material is toxic and inhalation should be avoided or limited. Further discussion of this concept is provided in *Academic Laboratory Chemical Hazards Guidebook*, Chap. 2.

Basically, the function of the laboratory hood or any other local ventilation equipment is to capture gases, vapors, and dusts so that a person working in front of the hood will not breathe a concentration higher than the TLV. If this is achieved, the concentration in the room will also be below the TLV.

Flammable Solvents

Many materials can form flammable or even explosive mixtures with air, but a dangerous concentration can be prevented by effective local ventilation. Work with flammable solvents should be performed in a hood so that the vapors will be not only captured but also diluted with air to a harmless concentration in the exhaust system.

Since the lower flammable limit is always much higher than the TLV, it follows that ventilation controlling the toxic hazard will always preclude a flammability hazard. However, a solvent spill, spreading volatile liquid over a substantial table area, can rapidly evaporate, forming a large volume of vapor that must be carried away and may cause the ventilation system to be overtaxed. To reduce the potential for a dangerous spill, the amounts of flammables in use should be kept to a minimum and small containers required. If work with volatile solvents is carried out above a safety pan (of metal, plastic, or another impervious material), the spread of vaporizing liquid will be limited in case of a spill or vessel breakage.

GENERAL LABORATORY VENTILATION

General ventilation refers to the quantity and quality of air supplied to the laboratory. The overall ventilation system should ensure that the laboratory air is continuously being replaced so that concentrations of odoriferous or toxic substances do not increase during the work day. Provided that auxiliary local exhaust systems are available and are used as the primary method for controlling concentrations of airborne substances, a ventilation system that changes the room air 4–12 times per hour is normally adequate.

In all cases, the movement of air in the general ventilation system for a building should be from the offices, corridors, etc. into the laboratories. All air from laboratories should be exhausted outdoors and not recycled. Thus, the air pressure in the laboratories should always be negative with respect to the rest of the building. The air intakes for a laboratory building should be in a location that reduces the possibility that the input air will be contaminated by the exhaust air from either the same building or any other nearby laboratory building. One common arrangement is to locate all of the laboratory-hood exhaust vents (the usual exhaust ports for laboratories) on the roof of the laboratory building and the building air-intake port at a different site where local movement is unlikely to mix exhaust air and intake air.

The laminar (i.e., nonturbulent) flow of incoming air is ideal and can be achieved by using a plenum or several louvers at the air-input sites of the laboratory. Air entry through perforated ceiling panels has also been used successfully to provide uniform airflow. The plenum, louver, or perforated ceiling panels should be designed so as to direct clean, incoming air over the laboratory personnel and sweep contaminated air away from their breathing zone. The size of the room and its geometry or configuration as well as the velocity and volume of input air will affect the room air patterns. However, it is difficult to offer generalizations about the effects of air-input and air-output ports on general laboratory ventilation.

Evaluation of General Laboratory Ventilation

Each laboratory should be evaluated in terms of the quality and quantity of its general ventilation. This evaluation should be repeated periodically and any time a change is made, either in the general ventilation system for the building or in some aspect of the local ventilation within the laboratory. This evaluation should begin by observing the pattern of air movement entering and within the laboratory.

Airflow paths into and within a room can be determined by observing smoke patterns. Convenient smoke sources are commercial smoke tubes available from local safety and laboratory supply houses. Aerosol generators that produce continuous, voluminous fogs are also available from the same sources. Such aerosol generators are generally used in ventilation research and for evaluating local exhaust ventilation, such as laboratory hoods. If the general laboratory ventilation is satisfactory, the movement of air from the corridors and other input ports through the laboratory to the hoods or other exhaust ports should be relatively uniform. There should be no areas where the air remains static or areas that have unusually high airflow velocities. If areas with little or no air movement are found, the ventilation engineer should be consulted and appropriate changes should be made in input or output ports to correct the deficiencies. Alternatively, signs warning of inadequate ventilation should be posted in such areas.

At low concentrations, most chemical vapors and gases tend to rise with warm air currents and become diluted with the general room air. Air movement in large laboratories is normally multidirectional and typically has a velocity of about 20 linear ft/min (lfm). This diverse air movement is a result of the movement of people and the effects of air intakes and exhausts and of eddy currents around benches and other fixed objects. The net result is that the general air composition is rather uniform. This mixing of room air does not mean that isolated static air spaces cannot be found, but rather that the general area occupied by the laboratory workers tends to be uniform in composition unless there are serious deficiencies in the locations of input and output ports.

The average time required for a ventilation system to change the air within a laboratory can be estimated from the total volume of the laboratory (usually measured in cubic feet) and the rate at which input air is introduced or exhaust air removed [usually measured in cubic feet/minute (cfm)]. The latter value is usually determined by measuring [usually in (lfm)] the average face velocity for each laboratory exhaust port, such as the hoods or other local ventilation systems. For each exhaust port, the product of the face area (in square feet) and the average face velocity (in lfm) will give the rate at which air is being exhausted by that port (in cfm). The sum of these rates for all exhaust ports in the laboratory will yield the total rate at which air is being exhausted from the laboratory. It is important to realize that, up to the capacity of the exhaust system, the rate at which air is exhausted from the laboratory will equal the rate at which input air is introduced. Thus, decreasing the flow rate of input air (perhaps to conserve energy) will decrease the number of air changes per hour in the laboratory, the face velocities of the hoods, and the capture velocities of all other local ventilation systems.

The measurement of airflow rates requires special instruments and personnel trained to use them. Pitot tubes are used for measuring duct velocities, and anemometers or velometers to measure airflow rates within rooms and at the faces of input or exhaust ports. These instruments are available from safety supply companies or laboratory supply houses. The proper calibration and use of these instruments and the evaluation of the data are a separate discipline; consultation with an industrial hygienist or a ventilation engineer is recommended whenever serious ventilation problems are suspected or when decisions on appropriate changes in the ventilation system to achieve a proper balance of input and exhaust air must be made.

Evaluation of Airborne Contaminants by Using TLVs

It is possible to calculate the amount of a chemical that can be emitted into the general laboratory atmosphere without exceeding the TLV or acceptable-exposure value. The air-saturation level (ASL) (in parts per million) can be calculated from the vapor pressure P (in mm Hg) by using Eq. (1):

$$\mathrm{ASL} = \frac{10^6(P)}{760} \qquad (1)$$

The results for some typical chemicals are given in Table 9-1. Most of these saturation levels are greater than the established TLVs, and some may even be life-threatening. For this reason, protective apparel including self-contained breathing apparatus (see Chap. 10) should be used to clean up major spills of highly toxic volatile chemicals because concentrations approaching saturation may be possible. Such chemical spills may also lead to airborne concentrations of chemicals in the explosive range (see Chap. 12), so that care should be taken to avoid any ignition source during cleanup operations.

Another illustrative calculation that can be made is the amount of a chemical that can be volatilized into the general laboratory atmosphere without exceeding the TLV, assuming a static atmosphere exists in the laboratory. For W grams of a substance of molecular weight M and a laboratory that has a volume V (in cubic feet), the concentration C (in ppm) at 25°C is given by Eq. (2):

$$C = \frac{8.65 \times 10^5(W)}{(V)(M)} \qquad (2)$$

Thus, if 454 g (1 lb) of acetone (MW 58.09) is spilled and allowed to evaporate in a laboratory that has a volume of 1000 ft^3, the concentration of acetone will be 6770 ppm, which substantially exceeds the TLV of

TABLE 9-1. Air-Saturation Concentration of Common Solvents

Solvent	Vapor Pressure at 20°C (mm Hg)	Saturation Level at 20°C (ppm)	ACGIH TLV (ppm)	OSHA PEL (ppm)
Acetone	184.8	2.43×10^5	750	1000
Benzene	74.2	9.76×10^4	10	1
Carbon tetrachloride	92.0	1.21×10^5	5	10
Chloroform	160.0	2.11×10^5	10	50
Diethyl ether	430.0	5.66×10^5	400	400
Dioxane	30.0	3.95×10^4	25	100
Ethanol	43.0	5.66×10^4	1000	1000
Ethylene glycol	0.05	65.8	50	
Hexane	119.0	1.57×10^5	50	500
Methylchloroform	100.0	1.32×10^5	350	350
Methylene chloride	349.0	4.59×10^5	50	500
Toluene	22.0	2.89×10^4	100	200

Source: From National Research Council, *Prudent Practices for Handling Hazardous Chemicals in Laboratories,* with permission from National Academy Press, Washington, D.C., 1981, p. 197.

750. The maximum amount of acetone that can be released into a laboratory of this size without exceeding the TLV is 50 g. For a more toxic material, such as benzene (MW 78.12, OSHA PEL = 1 ppm), the maximum permissible amount of material than can be released into a laboratory having a volume of 1000 ft^3 is about 0.1 g. The corresponding values for several common solvents are given in Table 9-2.

Such calculations illustrate the basis for the ACGIH recommendation that auxiliary local ventilation be used when working with a substance having a TLV of 50 ppm or less. For many such substances, 5 g or less per 1000 ft^3 of laboratory atmosphere will exceed the TLV.

More accurate estimations of general laboratory airborne concentrations of chemicals can be made if the dilution resulting from continuous air change is also considered. By assuming rapid diffusion and mixing, the general airborne concentration of a substance can be calculated by estimating its total emission into the room and dividing this by the rate of air exhaust from the room. For typical laboratories where work with a variety of different substances is in progress, additional calculations of this type are normally not warranted. If a reliable measure of the exposure to a specific substance is required, it is better obtained by having a laboratory worker wear a portable air-sampling device, packed with a suitable absorbent and mounted in the breathing zone, for a fixed time period. Analysis of the content of the sampling device provides a more accurate measure of the actual concentration of a specific substance to which the worker

TABLE 9-2. Amount of Vaporized Solvent per 1000 ft^3 of Air Required to Equal the TLV at 25°C

Solvent	MW	TLV (ppm)	Limit (g)	STEL (ppm)	Limit (g)
Acetone	58	750	50	1000	67
Benzene^{A2}	78	10	0.9		
Carbon tetrachloride^{A2}	154	5	0.9		
Chloroform^{A2}	119	10	50		
Diethyl ether	74	400	1.4	500	43
Dioxane	88	25	43		
Ethanol	46	1000	53		
Ethylene glycol	62	50^C	7		
n-Hexane	86	50	3.6		
Methylene chloride	85	50	4.9	500	49
Toluene	92	100	200	150	16
1,1,1-Trichloroethane	133	350	11	450	69

^{A2} denotes industrial substance suspected of human carcinogenic potential.
^{C} denotes a ceiling limit.
Source: From National Research Council, *Prudent Practices for Handling Hazardous Chemicals in Laboratories,* with permission from National Academy Press, Washington, D.C., 1981, p. 198.

has been exposed. Suitable air-sampling devices are available from various safety supply companies and laboratory supply houses.

Use of General Laboratory Ventilation

As discussed above, general laboratory ventilation is intended primarily to increase the comfort of laboratory workers and to provide a supply of air that will be exhausted by a variety of auxiliary local ventilation devices (hoods, vented canopies, vented storage cabinets, etc.). This ventilation provides only very modest protection from toxic gases, vapors, aerosols, and dusts, especially if they are released into the laboratory in any significant quantity. The cardinal rule for safety in working with toxic substances is that all work with these materials in a laboratory should be performed in such a way that they do not come into contact with the skin and that quantities of their vapors or dust that might produce adverse effects are prevented from entering the general laboratory atmosphere. Thus, operations such as running reactions, heating or evaporating solvents, and the transfer of chemicals from one container to another should normally be performed in a hood. If especially toxic or corrosive vapors will be evolved, these exit gases should be passed through scrubbers or adsorption trains. Toxic substances should be stored in cabinets fitted with auxiliary local ventilation, and laboratory apparatus that may discharge toxic vapors (vacuum pump exhausts, gas chromatograph exit ports, liquid chromatographs, distillation columns, etc.) should be vented to an auxiliary local exhaust system such as a canopy or snorkel. Samples that will be measured by using instruments or stored in apparatus where auxiliary local ventilation is not practical [such as balances, spectrometers, and refrigerators (see Chap. 5)] should be kept in closed containers during measurement or storage. Simply stated, laboratory workers should regard the general laboratory atmosphere only as a source of air to breathe and as a source of input air for auxiliary local ventilation systems.

Maintenance of Ventilation Systems

Even the best engineered and installed ventilation system requires routine maintenance. Blocked or plugged air intakes and exhausts may alter the performance of the total ventilation system. Belts loosen, bearings require lubrication, motors need attention, ducts corrode, and minor components fail: These malfunctions, individually or collectively, can affect overall ventilation performance.

All ventilation systems should have a monitoring device that readily permits the user to determine whether the total system and its essential components are functioning properly. Manometers, pressure gauges, and other devices that measure the static pressure in the air ducts are sometimes used to reduce the need for manually measuring the airflow. The need for and type of monitoring devices should be determined on a case-by-case basis. If the substance being contained has excellent warning properties and the consequence of overexposure is minimal, the system will need less stringent control than if the substance is highly toxic or has poor warning properties. The need for scheduled maintenance will also be determined by these factors.

LABORATORY HOODS

A chemical laboratory requires a ventilation system with fume hoods, as well as a general heating/cooling system. A makeup air system may also be necessary to replace the air exhausted by the hoods.

The function of the fume hoods is to capture, contain, and expel emissions generated by operations carried out in the hoods. The heating/cooling and makeup systems must be capable of providing enough air changes to dilute the concentrations of hazardous materials to a safe level.

Hoods are evaluated in terms of "face velocity." Face velocity is the average velocity of the air in feet per minute (fpm) in a direction perpendicular to the plane of the hood opening. Face velocities of 100–150 fpm are recommended, depending on the use of the hood for routine or hazardous operations. Routine use involves chemicals with low toxicities or small volumes of moderately toxic materials. A low-toxicity compound is defined as one having a TLV (TWA) of 500 ppm or greater and moderately toxic materials have a TLV (TWA) of 100–500 ppm. Hazardous operations involve large quantities of moderately toxic materials, any quantity of material with a TLV (TWA) of 100 ppm or less, any carcinogen, or any pathogenic organism.

Specific hood designs are available for many different hazards, including corrosive materials, flammable materials, and explosives. Hoods should be used for their intended purposes. Their success also depends on unrestricted airflow and their placement in the laboratory.

Fume hoods are not designed for the storage of chemicals. Hoods may not function properly when they are too close to doors, windows, and other restricting physical features. Airflow is impaired when the motor is weak, fan blades have been eaten away by corrosive gases and vapors, makeup air is not available in sufficient quantity, and ducts are clogged.

Hoods must be used with caution. Fire can be sucked into the ductwork. The window of a hood can be shattered by an explosion. Potential dangers must be identified in order to select the best hood.

The laboratory hood, usually purchased from a manufacturer of laboratory furniture, is the enclosure that is expected to capture all gases or aerosols released by materials placed within it or operations performed in it. Capture depends on face velocity, but can also be affected by the hood's location in the room and other external factors such as people passing by the hood. The face velocity or average speed of the air entering the open face of the hood should meet the following criteria:

Low to moderate hazard	100 ft/min
Moderate to high hazard	150 ft/min

Table 9-3 lists the various face velocities specified by different organizations for routine and hazardous operations involving hoods.

Face velocity is related to blower capacity by the formula:

$$\text{Face velocity (ft/min)} = \frac{\text{Capacity of blower (ft}^3\text{/min)}}{\text{Hood face opening (ft}^2\text{)}}$$

Thus, if a higher face velocity is needed, the hood sash should be partially closed. In fact, with a velometer to measure face velocity, it is feasible to mark off sash positions for various face velocities. If readings are consistent but too low, a stop may be installed on the sash track to prevent it from opening too wide. If a hood does not have a movable front sash, a sheet of plywood or rigid plastic may be hung on one side to reduce the area of the opening.

Hood efficiency is also dependent on placement in the laboratory. If a hood is located close to a door, drafts will often cause serious problems with airflow patterns in the hood. Even the disturbances in airflow patterns resulting from people walking past the front of the fume hood can reduce the efficiency of the hood.

Hood Design and Construction

Laboratory hoods and their exhaust ducts should be constructed of nonflammable materials. They should be equipped with either vertical or horizontal sashes that can be closed. Welded steel construction is recommended for the sash frame. The glass within the sash should be laminated safety glass at least 7/32 in. thick or of another equally safe material that will not shatter in the event of an explosion within the hood. The utility control valves, electrical receptacles, and other fixtures should be located outside of the hood to minimize the need to reach within the hood itself. The construction materials, plumbing requirements, and interior design will vary, depending on the intended use of the hood. Although external baffles have been shown to provide improved directional airflow on some hoods, their use on all hoods is not indicated.

In recent years, the "supplementary-air hood" has become popular. This hood directs a blanket of unconditioned or partially conditioned air vertical to the hood face between the operator and the sash. As much as 70% of the total air exhausted by the hood can be taken from a supplementary air source, resulting in considerable savings in energy. However, the careful balancing of both the velocity and direction of incoming air is required to achieve even air distribution across the hood face; if such a hood is operated with the sash closed, the supplementary air is of little value and may even upset the general laboratory ventilation. Consequently, the design and installation of a laboratory ventilation system employing supplementary-air hoods should not be attempted without the aid of a qualified ventilation engineer.

TABLE 9-3. Recommended Minimum Average Hood Face Velocities in Feet per Minute (fpm)

Organization	Routine Operation	Hazardous Operations
American Chemical Society (ACS)	100	125–200
American Conference of Government Industrial Hygienists (ACGIH)	100	150
American National Standards Institute (ANSI)	100	150
U.S. National Institute for Occupational Safety and Health (NIOSH)	100	150

Source: From NIOSH, *Safety in the School Science Laboratory, Instructors Resource Guide,* 1979, p. 11-16.

Although hoods are most commonly considered as devices for controlling concentrations of toxic vapors, they can also serve to dilute and exhaust flammable vapors. Almost any hood can be used effectively or it can be overloaded or misused, resulting in spillage of vapor into the general room air. Also, an overloaded hood may contain an explosive mixture of air and a flammable vapor. Both the hood designer and user should be cognizant of this hazard and eliminate possible sources of ignition within the hood and its duct work if potential for explosion exists. The concentration of vapors in the hood atmosphere can be estimated by using the procedures described above for calculating general room concentrations.

In some cases, the materials that might be exhausted by a hood are sufficiently toxic that they cannot be expelled into the air. Whenever possible, experiments involving such materials should be designed so that the toxic materials are collected in traps or scrubbers rather than being released into the hood. If, for some reason this is impossible, HEPA (high-efficiency particulate filter) filters are recommended for highly toxic particulates and activated charcoal filters can be used to adsorb highly toxic gases and vapors. Liquid scrubbers may also be used to remove both particulates and vapors and gases. None of these methods is completely effective. Incineration ultimately may be the best method for destroying combustible compounds in exhaust air, but adequate temperature and dwell time are required to ensure complete combustion. Incinerators require considerable energy, and other methods should be studied before resorting to their use. The optimum system for collecting or destroying toxic materials in exhaust air must be determined on a case-by-case basis. In all cases, such treatment of exhaust air should be considered only if it is not practical to pass the gases or vapors through a scrubber or adsorption train before they can enter the stream of hood exhaust air.

Hood Blowers

Each hood is preferably served by its own blower, properly selected for delivery and pressure drop characteristics. All blowers should be located outside the building, usually on the roof, so that all interior ductwork is under negative pressure. Discharge from the blower(s) should be directed upward with no obstruction such as a weather cap. Discharge locations should be fairly remote from air conditioning equipment or other air intake so as to avoid recycling of the hood exhaust.

Motors driving exhaust blowers are usually outside of the air stream; hence, they are not required to be explosion-proof but should be weather-tight. Each blower motor should be controlled by a switch at the front of the hood and a pilot light should indicate when the hood is on.

Ductwork

Ductwork for laboratory hoods should be of a material resistant to chemicals handled in the hoods. Fiberglass, asbestos cement, and steel (sometimes rubber-coated) are common choices. Joints should be tightly made and cemented so they do not leak. Sound deadening may be needed in metal hoods to reduce noise.

The ducts should be dedicated to that purpose and not combined with other ventilation ducts within the building. It is usually best to have a separate duct for each hood to eliminate the possibility that toxic vapors exhausted into one hood could be channeled into an unused hood and reenter the general laboratory atmosphere. The design of the manifold is rather complex and the proper balance of air volume is difficult. If several hoods are connected to a common exhaust duct, then some fail-safe arrangement should be provided to ensure that all of them are continually exhausting air when any one of them is in use.

Adding another hood to an existing duct system usually means that one or more hoods will have inadequate face velocity.

USE OF LABORATORY HOODS

Although many laboratory workers regard hoods strictly as local ventilation devices to be used to prevent toxic, offensive, or flammable vapors from entering the general laboratory atmosphere, hoods offer two other significant types of protection. Placing a reacting chemical system within a hood, especially with the hood sash closed, also places a physical barrier between the workers in the laboratory and the chemical reaction. This barrier can afford the laboratory workers significant protection from hazards such as chemical splashes or sprays, fires, and minor explosions. Furthermore, the hood can provide an effective containment device for accidental spills of chemicals. In a laboratory where workers spend most of their time working with chemicals, there should be at least one hood for every two workers, and the hoods should be large enough to provide each worker with at least 2.5 linear ft of working space at the face. The optimum arrangement is to provide each laboratory worker with a separate hood. In circumstances where this amount of hood space cannot be provided, there should be reasonable provisions for other types of local ventilation, and special care should be exercised in monitoring and restricting the use of hazardous substances.

The following factors should be remembered in the daily use of hoods:

1. Hoods should be considered backup safety devices that can contain and exhaust toxic, flammable, or offensive materials when the design of an experiment fails and vapors or dusts escape from the apparatus being used. Hoods should not be regarded as the means for disposing of chemicals. Thus, apparatus used in hoods should be fitted with condensers, traps, or scrubbers to contain and collect waste solvents or toxic vapors or dusts. Highly toxic or offensive vapors should always be scrubbed or adsorbed before the exit gases are released into the hood exhaust system.
2. Hoods should be evaluated before use to ensure adequate face velocities (typically 60–100 lfm) and the absence of excessive turbulence. Furthermore, some continuous monitoring device for adequate hood performance should be present and checked before each hood is used. If inadequate hood performance is suspected, it should be established that the hood is performing adequately before it is used.
3. Except when adjustments of apparatus within the hood are being made, the hood should be kept closed: vertical sashes down and horizontal sashes closed. Sliding sashes should not be removed from horizontal sliding-sash hoods. Keeping the face opening of the hood small improves the overall performance of the hood.
4. The airflow pattern, and thus the performance of a hood, depend on such factors as placement of equipment in the hood, room drafts from open doors or windows, persons walking by, or even the presence of the user in front of the hood. For example, the placement of equipment in the hood can have a dramatic effect on its performance. Moving an apparatus 2–4 in. back from the front edge into the hood can reduce the vapor concentration at the face by 90%.
5. Hoods are not intended primarily for the storage of chemicals. Materials stored in them should be kept to a minimum. Stored chemicals should not block vents or alter airflow patterns. Whenever practical, chemicals should be moved from hoods to vented cabinets for storage.
6. Solid objects and materials (such as paper) should not be permitted to enter the exhaust ducts of hoods as they can lodge in the ducts or fans and adversely affect their operation.
7. An emergency plan should always be prepared for the event of ventilation failure (power failure, e.g.) or other unexpected occurrences such as fire or explosion in the hood.
8. If laboratory workers are certain that adequate general laboratory ventilation will be maintained when the hoods are not running, hoods not in use should be turned off to conserve energy. If any doubt exists, however, or if toxic substances are being stored in the hood, the hood should be left on. Energy can also be conserved by the use of variable-volume hoods that modulate exhaust flow with the sash position.

Hood Misuse

The laboratory hood is too often misused by storing toxic, flammable, or odorous materials in it. This should be avoided since it prevents the proper use of the hood and may obstruct the proper flow of air through the hood. Furthermore, when the blower is not running, no vapor is being removed and no protection exists. If a fire should occur in a cluttered hood, extinguishment may be difficult or impossible.

General-purpose laboratory hoods should not be used for specialized operations such as the use of perchloric acid, highly toxic materials, pathogenic organisms, or radioactive isotopes. Special designs are required for each such use.

HOOD INSPECTION AND MAINTENANCE

The hood is the best-known local exhaust device used in laboratories. It is, however, but one part (typically, the principal exhaust port) of the total ventilation system and should not be considered as separate from the total system, because its performance will be strongly influenced by other features in the general ventilation system. A number of recent studies of hood performance in terms of the protection hoods afford laboratory workers have shown the importance of such factors as the volume of input air to the laboratory, the location of the laboratory input-air ports, the location of the hood within the laboratory, and the placement of apparatus within the hood. Any efforts to correct poor hood performance should involve the consideration of all these factors.

Periodic testing and maintenance are vital to the continued effectiveness of a ventilation system. Great disparities among results across the hood face indicate that plenum slots need adjustment, the air passages need cleaning, or other factors have upset the draft balance. Too much material or equipment in the hood can affect flow. Sometimes, blades are eaten away by corrosive gases or vapors. Regular inspection and lubrication are required.

Other causes of poor hood performance are plugged ductwork and corroded ductwork that has leaks in it. Horizontal duct runs are susceptible to plugging by dirt and dust. They should be inspected and cleaned regularly. The dirt cuts down the cross-sectional area of the duct and reduces the airflow. Ducts that have holes from corrosive acids will allow the infiltration of large quantities of air, which reduces the quantity of air pulled through the hoods. Corroded ductwork should be replaced immediately.

Air cannot be exhausted unless it can be replaced by makeup air. For example, if a hood is located in a tightly closed room, the blower's effectiveness will be significantly decreased due to the increased pressure differential, resulting from inadequate air infiltration (makeup air). Likewise, a hood sash should not cut off all entering air, and stops are often used to maintain a 2–3 in. slot under the sash.

EVALUATION OF HOOD PERFORMANCE

All hoods should be evaluated for performance when they are installed. New hood types or designs should be evaluated by a method that quantitatively rates hood performance.

Performance should be evaluated against the design specifications for uniform airflow across the hood face, as well as for the total exhaust air volume. Equally important is the evaluation of operator exposure or another purpose for prescribing the hood. This evaluation of hood performance should be repeated any time there is a change in any aspect of the ventilation system. Thus, changes in the total volume of input air, changes in the locations of air-input ports, or the addition of other auxiliary local ventilation devices (such as more hoods, vented cabinets, and snorkels) call for the reevaluation of the performance of all hoods in the laboratory.

Face Velocity

The first step in the evaluation of hood performance is the use of a smoke tube or similar device to determine that the hood is on and exhausting air. A smoke tube contains titanium tetrachloride absorbed on a granular medium. When air is passed through the tube, the moisture in the air reacts with the titanium tetrachloride to form hydrochloric acid smoke. This smoke can be irritating to the eyes, nasal passages, and skin. Smoke tubes are most useful in visualizing flow patterns.

The second step is to measure the velocity of the airflow at the face of the hood. Face velocity can be measured with any one of several types of direct-reading meters. The operator should stand to one side and hold the meter at arm's length to avoid distortions due to eddy currents around his or her body. Sets of measurements should be taken with the hood sash fully opened and with the sash in one or more partially closed positions.

The third step is to determine the uniformity of air delivery to the hood face by making a series of face velocity measurements taken in a grid pattern. The measuring instruments, anemometers or velometers, should be calibrated before use. For an average 4- to 6-ft-wide hood, velocity measurements should be made at nine locations: three across the middle, three across near the top, and three near the working surface. In the measurement of airflow velocities at various points across the hood face, values for specific points may vary by ±25% from the average value. Greater variation than this should be corrected by adjustments in the interior hood baffle or, if necessary, by altering the path of the input air flowing into the room. Most laboratory hoods are equipped with a baffle that has movable slot openings at both the top and bottom. This baffle should be moved until the airflow (measured with an anemometer or velometer) approaches a uniform level. Larger hoods may require additional slots in the baffle to achieve uniform airflow across the hood face. If the sash is closed, the airflow will be less even.

The total volume of air being exhausted by a hood is the product of the average face velocity and the area of the hood face opening. If the hood and general ventilating system are properly designed, face velocities in the range of 60–100 lfm will provide a laminar flow of air over the floor and sides of the hood. High face velocities (150 lfm or more), which exhaust the general laboratory air at a greater rate, are both wasteful of energy and likely to degrade hood performance by creating air turbulence at the hood face and within the hood. Such air turbulence can cause the vapors within the hood to spill out into the general laboratory atmosphere.

The rate of air exhaust from a hood is determined, in part, by the size and speed of the hood exhaust fan. In some cases, hoods are fitted with multiple-speed fan motors so that fan speed can be selected by using a switch on the hood face. Such controls, dampers or baffles, and other devices can be engineered into the hood to reduce the venting of conditioned air. However, in all cases, the designed air-exhaust rate of the hood will be achieved only if adequate input air is supplied to the laboratory. If the volume of input air is not sufficient, changes in the hood fan speed will do little to improve hood performance.

Recently, in an effort to conserve energy, many laboratories have decreased the amount of input air, turned off some of their hoods, or done both during certain times of the day. Before such procedures are initiated, the effects on both hood performance and the overall laboratory ventilation system should be tested to avoid serious problems from inadequate ventilation. For example, if the hoods are the only exhaust ports in a laboratory, when they are all turned off, there may be no change of air in the laboratory. Alternatively, if all hoods are left on but the supply of input air is decreased, the performance of every hood in the building may be degraded to a potentially dangerous level because none of the laboratories will have an adequate supply of input air. A hood that is not providing adequate ventilation performance is often worse than no hood at all because the laboratory worker is likely to have a false sense of security about the protection provided by the local ventilation system he or she is using. Assuming that the general ventilation system is properly designed, any decrease in the amount of input air being supplied should be accompanied by turning off selected hoods to maintain a balance between input and exhaust air.

The optimum face velocity of a hood (also called the capture velocity) will vary depending on its configuration. As noted above, too high a face velocity is likely to increase the turbulence within the hood and cause gases or vapors to spill from the hood into the room. It is necessary only that the capture velocity under use conditions be greater than cross currents of air at the hood face.

As the sash is closed, the hood face area is reduced and the capture velocity increased. Thus, a hood that has low airflow and low capture velocity when the sash is open may provide excellent exposure control when the sash is closed. Likewise, a hood that has adequate airflow but is operated with the sash open in a cross draft or another source of air turbulence may not provide adequate exposure control.

Air Turbulence

The second aspect of hood performance that should be evaluated is the presence or absence of air turbulence at the face of and within the hood. The observation of smoke patterns is used for this evaluation. Visible fumes or smoke can be generated by using cotton swabs dipped in titanium tetrachloride, commercial smoke tubes, or aerosol generators. Each hood should be tested for air turbulence and capture effectiveness before it is used and again after any change is made in the overall ventilation system of the laboratory. If there is excessive turbulence or the hood fails to capture smoke, changes may be required in the hood face velocity, the location of the air-input ports, the physical location of the hood, or the volume of input air.

The hood's location within the laboratory will affect its performance. If the hood is placed so that cross drafts from the movement of people or air currents from open windows or doors exceed the capture velocity, material may be drawn from the hood into the room. Often, air turbulence at the hood face is best diminished by relocating the air-input ports or adding external baffles near the hood face. A qualified ventilation engineer should be consulted for aid in solving such problems.

Blocked Airflow

Another factor that influences the performance of a hood is the amount and location of equipment in it. As in the general laboratory, the air in the hood moves in all directions. Hot plates, heating mantles, and equipment standing in the hood may alter this movement and increase air turbulence. If the emission source is placed near the hood face, the vapors are likely to spill outside the hood. Although a high capture or face velocity may help prevent this spillage, a more satisfactory (and less energy-consuming) approach is to place the equipment farther into the hood. All equipment should be placed at least 4 in. back from the hood sash; in general, all equipment should be placed as far to the back of the hood as practical. In some laboratories, a colored stripe is painted on the hood work surface 4 in. back from the face to serve as a constant reminder. Some hoods have a 4-in.-wide raised ridge at the front to prevent workers from placing equipment at the opening and also to contain liquid spills. The less apparatus and bottles the hood contains, the more likely that it will have laminar airflow across its working surface. Observing these simple precautions will often result in a significant improvement in hood performance at face velocities of 60–80 lfm. Because a substantial amount of energy is required to supply conditioned input air to even a small hood, the use of hoods for the storage of bottles of toxic or corrosive chemicals is a very wasteful practice (see *Academic Laboratory Chemical Hazards Guidebook,* Chaps. 1 and 2), which can also, as noted above, seriously impair the effectiveness of the hood as a local ventilation device. Thus, it is preferable to provide separate vented cabinets for the storage of toxic or corrosive chemicals. The amount of air exhausted by such cabinets is much less than that exhausted by a properly operating hood.

The position and movement of the hood user will also affect the performance of the hood. When a user stands in front of an open hood sash, considerable turbulence and eddy currents may occur near the face of the hood. Placing equipment far back in the hood and partially closing the hood sash will help minimize losses caused by this air turbulence.

Continuous Monitoring of Performance

After the face velocity of each hood has been measured (and the airflow balanced if necessary), each hood should be fitted with an inexpensive manometer or another pressure-measuring device (or a velocity-measuring device) to enable the user to determine that the hood is operating as it was when evaluated. This pressure-measuring device should be capable of measuring pressure differences in the range of 0.1–2.0 in. of water and should have the lower-pressure side connected to the duct just above the hood and the higher-pressure side open to the general laboratory atmosphere. Once such a device has been calibrated by measuring the hood face velocity, it will serve as a constant monitor of hood performance and can provide warning of the inadequate hood performance that might arise from future defects or changes in the overall laboratory ventilation. In many cases, the indicator light on the front of a hood next to the switch simply indicates whether or not the switch is in the on position and provides no information about whether or not the hood fan motor is actually running or the input-air supply is adequate.

Monitoring Worker Exposure

Perhaps the most meaningful (but also the most time-consuming) method of evaluating hood performance is to measure worker exposure while the hood is being used for its intended purpose. By using commercial air-sampling devices that can be worn by the hood user, worker exposure, both the excursion peak and time-weighted average, can be measured using standard industrial hygiene techniques. The criterion for evaluating the hood should be the desired performance; that is, does it contain vapors and gases at the desired worker exposure level, and a sufficient number of measurements should be made to define a statistically significant maximum exposure based on worst-case operating conditions. Direct-reading instruments are available for determining the short-term concentration excursions that may occur in laboratory hood use. Various low-toxicity chemicals, such Freon 11, sulfur hexafluoride, and certain detectable dyes, can also be used with direct-reading instrumentation to evaluate hood performance. However, for most purposes, the combination of a smoke test and a series of face velocity measurements will be sufficient to evaluate hood performance.

LOCAL EXHAUST SYSTEMS AND SPECIAL VENTILATION AREAS

The usual laboratory hood depends on the horizontal flow of a substantial volume of air to direct contaminated air away from the user. Although canopy hoods, snorkels, and similar devices that use vertical airflow are much less effective, they are useful in providing local ventilation above chromatographic and distillation equipment and various instruments (e.g., spectrometers) that cannot reasonably be placed in hoods. With good design, these vertical-airflow devices can be used to contain emissions of hazardous substances. Drop curtains or partial walls of plastic or metal to direct the airflow may make them more effective. However, without proper training, the users of these devices may find themselves in a contaminated air stream.

Whether the emission source is a vacuum-pump discharge vent, a gas chromatography exit port, or the top of a fractional distillation column, the local exhaust requirements are similar. The total airflow should be high enough to transport the volume of gases or vapors being emitted, and the capture velocity should be sufficient to collect the gases or vapors. The capture velocity is approximately 7.5% of the face velocity at a distance equal to the diameter of the local exhaust opening. Thus, a 3-in. snorkel or elephant trunk tube having a face velocity of 150 lfm will have a capture velocity of only 11 lfm at a distance of 3 in. from the opening. As the air-movement velocity in a typical room is 20 lfm, the capture of vapors emitted 3 in. from the snorkel will be incomplete. However, vapors emitted at distances of 2 in. or less from the snorkel opening will be captured completely. Thus, these canopy-hood or snorkel devices will be effective only when the source of vapor emission is placed very close to the opening of the local exhaust system or if supplementary plastic or metal curtains are used to direct the airflow into the opening of the local exhaust device.

Despite these limitations, these canopy-hood and snorkel systems can provide the useful and effective control of emissions of toxic vapors or dusts if they are installed and used correctly. It is not considered good practice to attach such devices to an existing hood duct; a separate exhaust duct should be provided. One very important consideration is the effect that such added local exhaust systems will have on the remainder of the laboratory ventilation. Each snorkel or canopy hood added will be a new exhaust port in the laboratory and will compete with the existing exhaust ports for input air. Thus, before any extra local exhaust systems are added, a qualified ventilation engineer should be consulted. After such devices have been installed, all aspects of the ventilation system should be evaluated so that necessary changes in the volume of input air and in the location of input-air ports can be made. Failure to follow these precautions could result in the serious degradation of all aspects of the ventilation system in the laboratory.

Downdraft ventilation has been used effectively to contain dusts and other dense particulates and high concentrations of heavy vapors that, because of their weight, tend to fall. Such systems require special engineering considerations to ensure that the particulates are transported in the air stream. Knock-outs or areas of low air velocity may be designed into such systems to permit the collection of larger particles before the exhaust air is filtered or scrubbed.

Glove Boxes and Isolation Rooms

Glove boxes are usually small units that have multiple ports in which arm-length rubber gloves are mounted, and the operator works through these. Construction materials vary widely, depending on their intended use. Clear plastic is frequently used because it allows visibility of the work area and is easily cleaned.

Glove boxes generally operate under negative pressure, so that any air leakage is into the box. If the material being used is sufficiently toxic to require the use of an isolation system, it is obvious that the exhaust air will require special treatment before release into the regular exhaust system. However, because these small units have a low airflow, scrubbing or adsorption (or both) can be accomplished with little difficulty. Some glove boxes operate under positive pressure. These boxes are commonly used for experiments in which protection from atmospheric moisture or oxygen is desired. If such glove boxes must be used with materials that present a high toxicity hazard, they should be thoroughly tested for leaks before each use and there should be a method of monitoring the integrity of the system (such as a shutoff valve or a pressure gauge designed into it).

Isolation rooms employ the same principles as glove boxes, except that the protected worker is within the unit. The unit itself operates under negative pressure, and the exhaust air requires special treatment before release. Many isolation units have a separate air supply to prevent cross-contamination. The workers enter the unit through clean rooms in which they remove their street clothes and don clean work clothes and other personal protective equipment, such as supplied-air respirators. They reverse this procedure when leaving the isolation unit by removing their work clothes in a dirty room, passing through a shower, and then entering a clean room where they don street clothes.

These isolation areas require considerable engineering, and the training of personnel is most important. They are, however, frequently used for handling regulated carcinogens, although other equally potent but nonregulated carcinogens are being handled in regular laboratories by using carefully planned and monitored procedures and by following adequate safety procedures and local ventilation techniques. Isolation units should be used only for highly toxic substances with physical properties that make control difficult or impossible by more conventional methods; other substances should be handled by general safety procedures.

Chemical Storerooms

Many schools have a chemical storeroom where a wide variety of potentially hazardous materials are kept. Often, this is an unventilated interior room. The odor level in a chemical storeroom can be an indication of the presence of flammable or toxic substances.

Exhaust ventilation should be installed in the corner most remote from the door. For a small storeroom of 1000 ft^3 (30 m^3), a blower of 200–300 ft^3/min rating should run continuously.

Environmental Rooms

Environmental rooms, either as refrigeration cold rooms or as warm rooms for the growth of organisms and cells, have the inherent property (as a result of their construction) of being a closed air-circulation system. Thus, the release of any toxic substance in these areas poses potential dangers. Also, because of the contained atmosphere in these rooms, significant potential exists for the creation of aerosols and for cross-contamination of research projects. These problems should be controlled by preventing the release of aerosols or gases into the room environment.

Because of the contained atmosphere in environmental rooms, provisions should be made to allow persons in them to escape rapidly if necessary. The doors of these rooms should be equipped with magnetic latches (preferable) or breakaway handles that would allow a trapped person to dismantle the door; the electrical system should be independent of the main power supply so that such persons are not confined in the dark in the event of a general power outage.

As for other refrigerators, volatile flammable solvents should not be used in cold rooms. The exposed motors for the circulating fans can serve as a source of ignition and initiate an explosion. The use of volatile acids should also be avoided in these rooms because such acids can corrode the cooling coils in the refrigeration system, which can lead to the development of leaks of toxic refrigerants.

PERSONAL SAMPLING

The atmosphere of the chemical laboratory should be monitored regularly for prospective hazardous material. Samples of air should be taken with professional sampling equipment and then analyzed by standard techniques in a qualified laboratory.

For sampling, air is pumped through a filter, adsorbent or absorbent. The sample may be collected on a filter cassette, adsorbed on charcoal or silica gel and desorbed by carbon disulfide, or collected by absorption in a solution in a midgit impinger. The pumps are powered by rechargeable nickel cadmium batteries, safe for use in explosive atmospheres. They must be calibrated prior to each use. Pumps range in capacity, with the 50–200 mL/min and 1–2 L/min volumes being most common.

Samples are usually analyzed by spectrophotometry, HPLC, and gas chromatography methods. Silica gel and charcoal samples are usually desorbed by carbon disulfide and analyzed by gas chromatography.

SUGGESTED READINGS

American Conference of Governmental Industrial Hygienists, *Industrial Ventilation,* published biennially by the ACGIHs, Cincinnati, Ohio. An excellent and inexpensive source of ventilation design information.

American Conference of Governmental Industrial Hygienists, *Threshold Limit Values for Chemical Substances and Physical Agents in the Work Environment and Biological Exposure Indices with Intended Changes,* Cincinnati, Ohio. Use the latest edition.

10

PROTECTIVE AND EMERGENCY EQUIPMENT

INTRODUCTION

Personal protective equipment is one of the major defenses against hazards in the work environment. These devices should never be a substitute for good engineering in the design and operation of a laboratory. However, there are times and places when engineering cannot solve all the problems, at least given the present state of the art.

A variety of specialized clothing and equipment is commercially available for use in the laboratory. The proper use of these items will minimize or eliminate exposure to the hazards associated with many laboratory operations. The primary goal of laboratory safety procedures is the prevention of accidents and emergencies. However, accidents and emergencies may nonetheless occur and, at such times, proper safety equipment and correct emergency procedures can help minimize injuries or damage. Every laboratory worker should be familiar with the location and proper use of the available protective apparel and saftey equipment and with emergency procedures. Instruction on the proper use of such equipment, emergency procedures, and first aid should be available to everyone who might need it.

EYE AND FACE PROTECTION

The eyes are the most easily injured organs of the body. Eyes may be permanently damaged by acids, alkalies, flammable substances, and other corrosive chemicals. However, most eye injuries result from flying debris such as dust, exploded apparatus, and grinding chips.

Fortunately, the eyes can be protected better than any other part of the body. Safety glasses, goggles, and shields are the primary means of personal protection.

Most states have enacted legislation that requires the wearing of protective eye wear or face shields by students working in school science laboratories. Goggles serve to provide the wearer with minimum protection against solid particles or liquid droplets that could enter the eye.

Explosions, dropped glassware, and other accidents can all serve as the source of flying particles. Industrial-quality eye protection must be worn at all times by all people in the laboratory. It does not matter that the wearer might be engaged in a relatively innocuous task; as long as he is in the laboratory, he must wear appropriate eye protection.

A clear, firm eye protection policy should be adopted. Enforcement must necessarily be the teacher's function. A basic tenet should be "approved eye protection must be worn at all times by anyone in the laboratory." If the plan instructs students to wear it only when needed, the whole concept will break down. The policy should follow the state law and regulations or, in states not having a school eye protection act, the National Society for the Prevention of Blindness (NSPB) model law should give guidance. ANSI Z87.1 should be followed for the selection and purchase of eye protective equipment.

Approved eye protection should be worn when participating in or observing the following courses or activities using the following materials:

1. Chemical, physical, or combined chemical-physical laboratories involving caustic or explosive materials, hot liquids or solids, injurious radiation, or other hazards.
2. Vocational, technical, industrial arts, chemical, or chemical-physical courses, involving exposure to:
   (a) Hot molten metals or other molten materials
   (b) Milling, sawing, turning, shaping, cutting, grinding, or stamping of any solid materials
   (c) Heat treatment, tempering, or kiln firing of any metal or other materials

(d) Gas or electric arc welding, or other forms of welding processes
(e) Repair or servicing of any vehicle
(f) Caustic or explosive materials

Safety Glasses and Goggles

Safety glasses are usually made of plastic or special hardened glass. Lenses can be clear or colored to filter unwanted wavelengths of light. Some safety glasses have side shields to protect against splashes. Others have shielding all around. Headbands keep them from being moved or knocked off in case of accident.

Safety goggles are just variations of safety glasses. They fit tightly against the face and thus offer superior protection against liquids, vapors, and projectiles. Nothing is a better source of eye protection when hazardous operations are involved.

Contact lenses should never be worn in the lab, even with safety glasses or goggles. Chemicals can become trapped in the eye, causing serious damage before the lenses can be removed. Removal of the lenses, even by the wearer, is also a problem in case of accidents. If prescription glasses are required, they should be made of safety glass.

Safety Glasses. Ordinary prescription glasses do not provide adequate protection from injury to the eyes. Minimum acceptable eye protection requires the use of hardened glass or plastic safety spectacles.

Safety glasses used in the laboratory should comply with the Standard for Occupational and Educational Eye and Face Protection (Z87.1) established by ANSI. This standard specifies a minimum lens thickness of 3 mm, impact resistance requirements, passage of a flammability test, and lens-retaining frames.

Side shields attached to regular safety spectacles offer some protection from objects that approach from the side but do not provide adequate protection from splashes. Other eye protection, such as goggles, should be worn when a significant splash hazard exists.

Safety Goggles. Goggles are not intended for general use. They are intended for wear when there is danger of splashing chemicals or flying particles. For example, goggles should be worn when working with glassware under reduced or elevated pressure and when glass apparatus is used in combustion or other high-temperature operations. Impact-protection goggles have screened areas on the sides to provide ventilation and reduce fogging of the lens and do not offer full protection against chemical splashes. Splash goggles ("acid goggles") or face shields that have splash-proof sides should be used when protection from a harmful chemical splash is needed.

Disinfection. Where more than one student must use the same goggles, it is advisable to clean and sterilize them between wearings so that infection will not be spread. Cleaning solutions containing quaternary ammonium compounds are effective but require thorough rinsing before drying.

Some schools have installed wall cabinets for sterilizing goggles with ultraviolet light. These cabinets are expensive but may be used if sterilization time is carefully controlled. Such a cabinet also has the advantage of providing storage for the goggles.

Face Shields

Goggles offer little protection to the face and neck. Face shields are curved sheets of plastic that protect the entire face of the wearer. They are commonly worn when using concentrated acids and alkalies and when working with unknown reactions. Full-face shields that protect the face and throat should always be worn when maximum protection from flying particles and harmful liquids is needed—for example, when a vacuum system (which may implode) is used or when conducting a reaction with the potential for mild explosions. For full protection, safety glasses should be worn with face shields. The metal-framed "nitrometer" mask offers greater protection for the head and throat from hazards such as flying glass or other light fragments. The instructor must establish rules governing the use of face shields.

Safety Shields

Safety shields are used to protect the face and upper body against possible explosions or splash hazards. The various types of safety shields are generally constructed of plexiglass or polycarbonate plastics. The shields are movable, self-supporting devices that can be placed between the worker or student and the experiment on which he or she is working. The worker's body and face are protected. The hands may remain the only vulnerable portion of the body when barrier shields are used. When combined with eye and face protection, with heavy gloves and sleeves, a safety shield of plexiglass can provide excellent protection from explosions and splatters.

Provided its opening is covered by closed doors, the conventional laboratory exhaust hood is a readily available built-in shield. However, a portable shield should also be used when manipulations are performed, particularly with hoods that have vertical-rising doors rather than horizontal-sliding sashes.

In general, dangerous laboratory equipment should be shielded on all sides so that there is no line-of-sight exposure of personnel to flying objects. Portable shields can be used to protect against hazards of limited severity (e.g., small splashes, heat, and fires). A portable shield, however, provides no protection at the sides or back of the equipment and many such shields are not sufficiently weighted and may topple toward the worker when there is a blast (perhaps hitting him or her and also permitting exposure to flying objects). A fixed shield that completely surrounds the experimental apparatus can afford protection against minor blast damage.

Methyl methacrylate, polycarbonate, polyvinyl chloride, and laminated safety plate glass are all satisfactory transparent shielding materials. When combustion is possible, the shielding material should be nonflammable or slow-burning; if it can withstand the working blast pressure, laminated safety plate glass may be the best material for such circumstances. When cost, transparency, high tensile strength, resistance to bending loads, impact strength, shatter resistance, and burning rate are considered, methyl methacrylate offers an excellent overall combination of shielding characteristics. Polycarbonate is much stronger and self-extinguishing after ignition but is readily attacked by organic solvents.

Contact Lenses

Contact lenses should not be worn in the science laboratory. It has been argued that contact lenses offer protection from damage by particles and chemicals. Nothing could be more erroneous. Gases and vapors can be concentrated under such lenses and cause permanent eye damage. Soft lenses can absorb solvent vapors even through face shields and, as a result, adhere to the eye. Persons attempting to irrigate the eyes of an unconscious victim may not be aware of the presence of contact lenses, thus reducing the effectiveness of such treatment.

An eye contaminated by a chemical splash should be irrigated with water until the material has been completely washed out. This usually takes about 15 min. If a contact lens is in the affected eye, the chemical may be drawn under the lens by capillary attraction where it cannot be reached by water washing. The lens must be removed to permit effective washing. Under traumatic conditions with pain and fear as impediments, it may be impossible for the victim or anyone else to remove the lens because of the involuntary spasm of the eyelid.

Clearly, the only answer is to prevent the possibility of such an occurrence. Contact lenses should be discouraged or prohibited in the school laboratory. Students should wear spectacles for correction, covered by chemical goggles. If contact lenses are medically necessary and corrective glasses cannot be substituted for them, the lens wearer should be identified as a precaution, should an accident occur.

There are some exceptional situations in which contact lenses must be worn for therapeutic reasons. Persons who must wear contact lenses should inform their laboratory supervisor so that satisfactory safety precautions can be devised.

Laser Eye Protection

Lasers are used in some school laboratories. Eye protection in their use depends more on the energy, wall absorption, and proper procedures than on absorptive lenses. Each laser wavelength requires a particular color and density lens. Hence, lenses must be properly matched to the source, otherwise, little or no protection is afforded. Even with the proper protective lenses, one must *never* look directly into a beam.

There are specific goggles and masks for protection against laser hazards and ultraviolet or other intense light sources, as well as glass-blowing goggles and welding masks and goggles. The laboratory supervisor should determine whether the task being performed requires specialized eye protection and insist on the use of such equipment if it is necessary. See Chap. 11 for detailed information on laser hazards.

Eyewash Devices

If all protective measures fail and a student or teacher gets a corrosive chemical in the eye, an eyewash device should be available for the immediate and thorough washing of the eye. Every second counts when lab accidents involve the eyes. At least one eyewash device should be in every science laboratory where chemicals are used. Speed is essential. With an alkali splash, the first 10–30 sec are critical. The first line of defense is the eyewash station and/or shower, depending on the nature of the accident.

Eyewash Fountains. Eyewash fountains may be required if the substance in use presents an eye hazard or in research or instructional laboratories where unknown hazards may be encountered. These fountains should be located close to safety showers so that, if necessary, the eyes can be washed while the body is showered.

An eyewash fountain is a permanently installed basin designed to provide a stream of water to gently wash the eye. Eyewash fountains come in a number of configurations, including free-standing units or as part of an eyewash fountain-safety shower combination. To use one, just remove your safety glasses or goggles, lean over, and flood the eyes with dual streams of water for a minimum of 15 min.

Eyewash fountains usually have dual faucets. The valve is provided with push bars, push plates, or a foot pedal so that the unit may be operated by hand or foot or by pushing a bar with the head. Many are equipped with stay-open ball valves so that the fountain user can use her hands to keep her eyes open while they are being irrigated.

An eyewash fountain should provide a soft stream or spray of aerated water for an extended period (15 min). The recommended water supply pressure for eyewash fountains is 25 lb/in.^{2} (psi), and the recommended water supply temperature is 110° F or less. Water warmer than 110° F is too hot for comfort. Cold water, even ice water, is usable for irrigation purposes.

Eyewash fountains should be located within 50 ft (15–30 sec away) from every laboratory work station. Lab workers should be very familiar with all locations in order to reach them quickly in a time of panic.

If fountains are not available, a hose (5–8 ft) with an aeration faucet can be used. The victim can lie down and have another person hold his eye open while he washes it. In an emergency, any source of gently flooding water is the best first aid.

In remote areas without a direct source of water, either plastic portable gravity-feed units or pressurized units (similar to fire extinguishers) are used. (See Figs. 10-1 and 10-2.)

Plastic gravity-feed eyewash stations utilize a plastic water tank large enough to provide a flow rate of about 0.5 gal/min for 15 min. This means that the minimum container size should be 7.5 gal. Twin nozzles are mounted at the base of the tank. The units are usually wall-mounted.

Portable eyewash fountains are available that consist of the hose and spray head discussed above and a pressurized water tank. These units approximate a typical fire extinguisher in size and weight. They should be used if hazardous work must be performed in the field or any area in which eyewash fountains are not available.

Other devices sometimes provided for cleansing the eyes and that are used occasionally are eyewash bottles and chemical burn stations. Most are very inadequate.

Eyewash bottle stations should be discouraged. A typical station consists of a wall holder or bracket and a liter bottle of boric acid or a buffer solution. The bottle is usually equipped with a cup that fits over the eye. Too little liquid is contained in such a bottle, and water alone is better for flushing the eyes.

FIGURE 10-1. Plastic portable gravity-feed eyewash unit. (Photo courtesy of Lab Safety Supply, Inc., Janesville, Wis.)

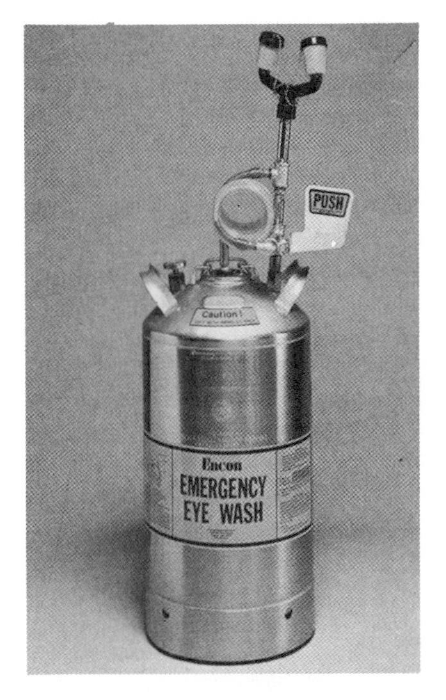

FIGURE 10-2. Pressurized eyewash unit.

Chemical burn stations consist of a wall bracket with a plastic bottle containing a phosphate buffer solution. The bottle's contents are poured into the eye.

The last two devices discussed, eyewash bottles and chemical burn stations, both have the disadvantages of being limited in cleansing capacity and difficult to find at a critical moment. They should never be considered as an alternative to eyewash fountains. Eye specialists favor water only and lots of it for 15 min minimum as the proper first-aid treatment for a chemical splash in the eye.

Important Note: Alkalies cause more eye damage than acids. Acids cause the eye to precipitate a protein substance that acts as a barrier to prevent penetration of the acid deeper into the eye. Caustic alkalies do not trigger any such protective response. Therefore, immediate eyewashing (within 30 sec) is very critical in the event of alkali burns.

Bacterial and Amoebic Contamination. Emergency eyewash stations are located in areas where caustic or corrosive chemicals are used. In facilities where plumbed-in water is available, regular unlimited-flow stations are used. In remote areas without a direct source of water, either plastic portable gravity-feed units or pressurized units (similar to fire extinguishers) are used (Figs. 10-1 and 10-2 previously illustrated the two types of units).

Amoebas can be present in emergency eyewash stations used by workers or students whose eyes have been traumatically exposed. Two genera of free-living soil amoebas are known to be potentially pathogenic to humans. Naegleria has been demonstrated as the causative agent for the often fatal disease of primary amoebic meningoencephalitis. Another genus, Acanthamoeba, has increasingly been found in amoebic keratitis, especially among contact lens users; such an infection usually results in irreversible eye damage, and victims often require the surgical treatments of corneal transplantation or eye enucleation.

Amoeba contamination has been reported in emergency eyewash stations, and it is recommended that all stations with a continuous water supply be flushed for 3 min weekly to keep the amoeba population to a minimum. Amoebas apparently cannot survive at elevated temperatures, such as those experienced by a plastic portable eyewash unit under the hot summer sun. Generally, 42–45°C is the critical limiting temperature for amoeba survival. It is likely, however, that amoebas can survive in these portable units during seasons when the water temperature does not reach this critical level. Increased pressure also appears to be a limiting factor in the survivability of amoeba. The stainless-steel fire-extinguisher-type eyewash stations are pressurized by compressed air or nitrogen to approximately 80 psi. Also, since these units were positively pressurized, foreign sources of contamination cannot readily enter the unit. The proper disinfection and subsequent refilling with distilled water of plastic portable units can eliminate organisms in the reservoir for at least three months. Due to the possible seriousness of the resulting disease and considering how easy it is to implement the suggested control measures, it is well worth the effort to conduct weekly flushing of the fixed units or to disinfect and maintain the plastic gravity-feed eyewash stations on a quarterly basis to assure that these units are free of amoebas. Users of the stainless-steel positively pressurized fire-extinguisher-type units may have already solved this problem.

Eye Protection Policy

Eye protection must be required for all personnel and any visitors present in locations where chemicals are stored or handled. No one should enter any laboratory without appropriate eye protection.

Conference rooms, libraries, offices, microscope rooms in which chemicals are not in use, and similar rooms are not normally eye protection areas. However, at any time when chemicals are used in such rooms, even temporarily, signs should be posted and all persons in the vicinity warned that eye protection is temporarily required. For laboratory operations that do not involve the use of chemicals or the use of chemicals in the immediate vicinity, it may be permissible by arrangement with the laboratory supervisor to remove eye protection.

Laboratory management should make appropriate eye protection devices available to visitors or others who only occasionally enter eye protection areas. These devices would be used only on a temporary basis while those persons are in the eye protection area. (For example, it may be useful to place a container of inexpensive safety glasses next to the entrance to each laboratory for use by visitors.)

BODY PROTECTION

In general, lab clothing should be comfortable but reasonably well-fitted. Tight garments restrict motion and hold spills in close contact with the body. Loose clothing is less controllable than tight-fitting clothing. Glassware can be knocked off benches, clothes can come into contact with an open flame, and manual dexterity can be reduced. Cotton offers better resistance to chemicals than wool or manmade fibers.

The clothing worn by laboratory workers can be important to their safety. Such personnel should not wear loose (e.g., saris, dangling neckties, and overlarge or ragged laboratory coats), skimpy (e.g., shorts and/or halter tops), or torn clothing and unrestrained long hair. Loose or torn clothing and unrestrained long hair can easily catch fire, dip into chemicals, or become ensnared in apparatus and moving machinery. Bare skin has no protection in the event of a chemical splash and street clothes offer minimal protection. If the possibility of chemical contamination exists, personal clothing that will be worn home should be covered by protective apparel.

Shoes should be worn at all times in buildings where chemicals are stored or used. Perforated shoes, sandals, or cloth sneakers should not be worn in laboratories or areas where mechanical work is being done.

Finger rings can react with chemicals and also should be avoided around equipment that has moving parts. Except for watches and plain rings, jewelry should be left at home or in the locker, along with scarves and other accessories.

Because the laboratory is a relatively hazardous environment, instructors and students should make every effort to reduce the chances of accident or exposure.

Protective Apparel

Appropriate protective apparel is advisable for most laboratory work and may be required for some. Such apparel can include laboratory coats and aprons, jump suits, special types of boots, shoe covers, and gauntlets. It can be either washable or disposable in nature. Garments are commercially available that can help protect the laboratory worker against chemical splashes or spills, heat, cold, moisture, and radiation.

Protective apparel should resist physical hazards and permit the easy execution of manual tasks while being worn. It should also satisfy other performance requirements: strength, chemical and thermal resistance, flexibility, and ease of cleaning. The required degree of performance can be determined on the basis of the substances being handled. The choice of garment—laboratory coat vs. rubber or plastic apron vs. disposable jump suit—depends on the degree of protection required and is the responsibility of the supervisor.

The traditional knee-length, long-sleeved lab coat of heavy cotton twill offers the best first-line protection against a variety of hazards. Twill stands up better than most other fabrics against chemicals, flying debris, and sharp edges. Fasteners allow the wearer to remove a coat quickly when spills occur.

Laboratory coats are intended to prevent contact with dirt and the minor chemical splashes or spills encountered in laboratory-scale work. The cloth laboratory coat may itself present a hazard (e.g., combustibility) to the wearer; cotton and synthetic materials such as Nomex or Tyvek[RT] are satisfactory; rayon and polyesters are not. Laboratory coats do not significantly resist penetration by organic liquids and, if significantly contaminated by them, should be removed immediately.

Plastic or rubber aprons provide additional protection when particularly hazardous or corrosive materials are being used. However, aprons do not take the place of a good lab coat. In labs where coats are not available, aprons should be worn over simple work clothes that cover the arms, legs, and body. Aprons and lab coats should be worn when working with corrosive or irritating reagents.

Plastic or rubber aprons can complicate injuries in the event of fire. Furthermore, a plastic apron can accumulate a considerable charge of static electricity and should be avoided in areas where flammable solvents or other materials could be ignited by a static discharge.

Disposable outer garments (e.g., Tyvek) may, in some cases, be preferable to reusable ones. One such case is that of handling appreciable quantities of known carcinogenic materials, for which long sleeves and the use of gloves are also recommended. Disposable full-length jump suits are strongly recommended for high-risk situations, which may also require the use of head and shoe covers. Many disposable garments, however, offer only limited protection from vapor penetration and considerable judgment is needed when using them. Impervious suits fully enclosing the body may be necessary in emergency situations.

Body suits and helmets are available when total exposure control is required. These suits are usually equipped with positive pressure to keep out dusts and vapors. Outside air sources are used.

Laboratory workers should know the appropriate techniques for removing protective apparel, especially any that has become contaminated. Chemical spills on leather clothing or accessories (watchbands, shoes, belts, and such) can be especially hazardous because many chemicals can be absorbed in the leather and then held close to the skin for long periods. Such items must be removed promptly and decontaminated or discarded to prevent the possibility of chemical burn.

Specialized or disposable clothing for use with particular classes of hazardous chemicals should be treated in a similar way.

Safety showers should be readily accessible for use when a chemical spill contaminates large sections of clothing or the skin.

Gloves

Gloves should be worn whenever it is necessary to handle corrosive materials, rough or sharp-edged objects, very hot or very cold materials, or whenever protection is needed against accidental exposure to chemicals. Gloves have a tendency to reduce dexterity, which may be a hazard in itself. Gloves should not be worn around moving machinery.

To prevent skin contact with toxic materials, it is important that the proper steps be taken to prevent such contact:

1. Proper protective gloves (and other protective clothing, when necessary) should be worn whenever the potential for contact with corrosive or toxic materials and materials of unknown toxicity exists.
2. Gloves should be selected on the basis of the material being handled, the particular hazard involved, and their suitability for the operation being conducted.
3. Before each use, gloves should be inspected for discoloration, punctures, and tears.
4. Before removal, gloves should be washed appropriately. (*Note:* Some gloves, e.g., leather and polyvinyl alcohol, are water-permeable.)
5. Glove materials are eventually permeated by chemicals. However, they can be used safely for limited time periods if specific use and glove characteristics (i.e., thickness and permeation rate and time) are known. Some of this information can be obtained from glove manufacturers, or the gloves used can be tested for breakthrough rates and times.
6. Gloves should be replaced periodically, depending on the frequency of use and permeability to the substance(s) handled. Gloves overtly contaminated (if impermeable to water) should be rinsed and then carefully removed.

Many different types of gloves are commercially available.

1. Leather gloves may be used for handling broken glassware, for inserting glass tubes into rubber stoppers, and for similar operations where protection from chemicals is not needed.
2. Cuffed, elbow-length rubber or plastic gloves protect the wearer from corrosive and hazardous liquids. There are various compositions and thicknesses of rubber gloves. Common glove materials include neoprene, polyvinyl chloride, nitrile, and butyl and natural rubbers. These materials differ in their resistance to various substances. Table 10-1 lists the chemical resistance of these glove materials to specific substances. More information on this topic is often available in glove manufacturers' catalogs. Rubber gloves should be inspected before each use; periodically, an inflation test, in which the glove is first inflated with air and then immersed in water and examined for the presence of air bubbles, should be conducted.
3. Insulated gloves should be used when working at temperature extremes. Various synthetic materials such as Nomex^{RT} and Kevlar^{RT} can be used briefly up to 1000°F. Gloves made with these materials or in combination with other materials such as leather are available. Fiberglass gloves also protect their wearers from heat. It is best not to use gloves made either entirely or partly of asbestos, which is regulated as a carcinogen by OSHA.
4. Specialized gloves are manufactured for electrical linesmen, welders, and others. It is the responsibility of the laboratory supervisor to determine whether specialized hand protection is needed for any operation and to ensure that the needed protection is available.
5. Cotton gloves are sometimes used to protect glassware from fingerprinting, and they are better than nothing as protection against broken fragments.

Footware

Shoes for the laboratory should completely cover the feet. Comfortable low-heeled loafers or tied shoes are recommended. High heels, platforms, and sandals are very dangerous and should be prohibited.

More extensive foot protection than ordinary shoes may be required in some cases. Rubber boots or plastic shoe covers may be used to avoid possible exposure of the feet to corrosive chemicals or large quantities of

TABLE 10-1. Resistance to Chemicals of Common Glove Materials (E = Excellent, G = Good, F = Fair, P = Poor)

Chemical	Natural Rubber	Neoprene	Nitrile	Vinyl
Acetaldehyde	G	G	E	G
Acetic acid	E	E	E	E
Acetone	G	G	G	F
Acrylonitrile	P	G	—	F
Ammonium hydroxide (sat.)	G	E	E	E
Aniline	F	G	E	G
Benzaldehyde	F	F	E	G
Benzene[a]	P	F	G	F
Benzyl chloride[a]	F	P	G	P
Bromine	G	G	—	G
Butane	P	E	—	P
Butyraldehyde	P	G	—	G
Calcium hypochlorite	P	G	G	G
Carbon disulfide	P	P	G	F
Carbon tetrachloride[a]	P	F	G	F
Chlorine	G	G	—	G
Chloroacetone	F	E	—	P
Chloroform[a]	P	F	G	P
Chromic acid	P	F	F	E
Cyclohexane	F	E	—	P
Dibenzyl ether	F	G	—	P
Dibutyl phthalate	F	G	—	P
Diethanolamine	F	E	—	E
Diethyl ether	F	G	E	P
Dimethyl sulfoxide[b]	—	—	—	—
Ethyl acetate	F	G	G	F
Ethylene dichloride[a]	P	F	G	P
Ethylene glycol	G	G	E	E
Ethylene trichloride[a]	P	P	—	P
Fluorine	G	G	—	G
Formaldehyde	G	E	E	E
Formic acid	G	E	E	E
Glycerol	G	G	E	E
Hexane	P	E	—	P
Hydrobromic acid (40%)	G	E	—	E
Hydrochloric acid (conc.)	G	G	G	E
Hydrofluoric acid (30%)	G	G	G	E
Hydrogen peroxide	G	G	G	E
Iodine	G	G	—	G
Methylamine	G	G	E	E
Methyl cellosolve	F	E	—	P
Methyl chloride[a]	P	E	—	P
Methyl ethyl ketone	F	G	G	P
Methylene chloride[a]	F	F	G	F
Monoethanolamine	F	E	—	E
Morpholine	F	E	—	E
Naphthalene[a]	G	G	E	G
Nitric acid (conc.)	P	P	P	G
Perchloric acid	F	G	F	E
Phenol	G	E	—	E
Phosphoric acid	G	E	—	E
Potassium hydroxide (sat.)	G	G	G	E
Propylene dichloride[a]	P	F	—	P
Sodium hydroxide	G	G	G	E
Sodium hypochlorite	G	P	F	G
Sulfuric acid (conc.)	G	G	F	G

See page 102 for footnotes.

TABLE 10-1. *(Continued)*

Chemical	Natural Rubber	Neoprene	Nitrile	Vinyl
Toluene[a]	P	F	G	F
Trichloroethylene[a]	P	F	G	F
Tricresyl phosphate	P	F	—	F
Triethanolamine	F	E	E	E
Trinitrotoluene	P	E	—	P

[a]Aromatic and halogenated hydrocarbons will attack all types of natural and synthetic glove materials. Should swelling occur, the user should change to fresh gloves and allow the swollen gloves to dry and return to normal.

[b]No data on the resistance to dimethyl sulfoxide of natural rubber, neoprene nitrile rubber, or vinyl materials are available; the manufacturer of the substance recommends the use of butyl rubber gloves.

Source: From National Research Council, with permission from *Prudent Practices for Handling Hazardous Chemicals in Laboratories*, National Academy Press, Washington, D.C., 1981, pp. 159–160.

solvents and water that might penetrate normal foot gear (e.g., during cleanup operations). Because these types of boots and covers may increase the risk of static spark, their use in normal laboratory operations is not advisable.

Other specialized tasks may require footwear that has, for example, conductive soles or insulated soles. Safety steel-toe shoes may be required in some parts of a laboratory or in other hazardous locations. The laboratory supervisor should recommend the use of such protection whenever appropriate.

Other Clothes

Hard hats are required in many industrial and construction sites. The use of a hard hat in laboratory environments is recommended especially where large equipment is operated.

The double locker system (one locker for street clothes and one locker for laboratory clothes) can be used in areas where reducing contamination is particularly critical.

Safety Showers

Safety showers should be provided in areas where chemicals are handled, for immediate first-aid treatment in the event of chemical splashes, spattering reactions, explosions, contact with toxic reagents and for extinguishing clothing fires. Their purpose is to provide a flood of water to the affected area. Every laboratory worker should learn the locations of and how to use the safety showers in the work area so that he or she can find them with eyes closed, if necessary. Safety showers should be tested routinely by laboratory personnel to ensure that the valve is operable and to remove any debris in the system.

The shower should be capable of drenching the subject immediately and should be large enough to accommodate more than one person if necessary. Safety shower heads must be of the nonclogging deluge type, thereby permitting the instantaneous flooding of a contaminated area of skin with water. The valve should be quick-opening requiring manual closing. To actuate the valve, a downward-pull delta-shaped bar is satisfactory if long enough, but chain pulls are not advisable because of the potential for persons to be hit by them and the difficulty of grasping them in an emergency.

Safety showers are commonly installed in corridors so that they can serve more than one laboratory. The number of showers to be installed and their locations must be given careful consideration. Large laboratory rooms may require the installation of one or more showers inside the room rather than in the corridor. Of course, safety showers should be immediately available in laboratory areas where the chance for harm is much greater. Safety showers should be installed in any laboratory where the danger of caustic or acid burns, contact with toxic chemical reagents, or clothing fires exists.

There are no uniform generally accepted requirements governing the installation, use, and maintenance of safety showers. The following specifications governing shower installation have been reported in the literature as acceptable for most laboratory situations.

1. Showers be located no more than 25 ft from the laboratory entrance. This specification obviously has to be modified by good judgment as mentioned above.
2. No individual in the laboratory should have to travel further than 50 ft to reach a shower.
3. The shower locations should be indicated by painted circles or squares on the floor. This area should be kept clear.
4. Showers should be located near floor drains if possible.
5. Shower heads should be located 7–8 ft above the floor and a minimum of 25 in. from the nearest wall.
6. The shower valve should be operated by a ring and chain, triangle and rod, or chain arrangement. The ring and triangle must be large enough to allow the entire hand to fit inside and grasp the ring or triangle comfortably. Safety shower valves are of two types: (1) a self-closing valve that remains open for about a minute each time it is opened, and (2) a full-flow ball valve that remains open until it is closed.
7. The shower flow rate should be between 30 and 60 gal/min and the water pressure should be 20–50 psi.
8. The water service line should be a minimum 1 in. in diameter.
9. Showers should not be located in the vicinity of electrical apparatus or power outlets or panels.

Each shower should have a shut-off valve protected from tempering or unauthorized persons. Shut-off valves should never be left closed, except while work is being done on the shower.

Safety Shower Tests. Showers should be tested at least every six months, but preferably more often to ensure that they are functioning properly. Plumbing fixtures corrode and some shower heads plug up. In areas where water hardness is high, the showers should be checked every month. Regular tests will reveal any problems that can then be corrected.

The shower test requires the use of a 55-gal drum mounted on four soft rubber wheels. The drum is equipped with a valve and drain pipe so that it may be drained after each shower is tested. A clear polyethylene containment sleeve is affixed around the shower head. The open end of the sleeve is inserted in the portable drum in such a manner that the discharged water is contained within the sleeve and the drum.

The test involves activating the shower valve and observing whether the valve opens and closes correctly, whether there are leaks, and whether the water is clear (free of rust and scale). From the quantity of water discharged, the adequacy of flush duration can be judged. A visual inspection of the installation is made. The actuating device is inspected for defects; the condition of the safety-shower symbol is noted; and the safety-shower location light, should the shower be equipped with one, is checked to see if it needs relamping.

A problem frequently encountered is the propensity of building occupants to hang the pull ring that activates the valve over the valve lever, thus rendering the device useless in the event of an emergency. Usually, this is done to prevent the pull ring from interfering with activities under the shower, and it obviously cannot be allowed. The lever and pull ring should be modified so that they will not swing out of reach.

HEARING PROTECTION

When high noise levels are a problem, the best solution is to reduce the noise level at the source by proper design and modifications. However, the wearing of devices to reduce noise levels is another solution. Air Force Regulation 161-35 allows noise levels for exposure times as follows:

Noise Level (dBA), hr	Exposure Time
84	8
88	4
92	2
96	1
100	0.5
104	0.25

These levels are more conservative than the limits enforced by OSHA.

Ear plugs and ear muffs are the basic kinds of hearing protection. Generally, ear muffs have a greater attenuation factor than ear plugs and, therefore, they are preferred.

Ear Plugs

Ear plugs are conical or cylindrical devices that seal the ear against the entrance of sound. The better the fit, the better the seal. Plugs can attenuate low-frequency sound by 25 dB and frequencies over 1000 Hz by 40 dB. However, these numbers can be substantially reduced by a poor fit.

Ear Muffs

Ear muffs resemble large headphones. The cups should be large enough to cover or seal off the ear. The muffs are filled with liquid, grease, or foam rubber cushions. Properly fitted muffs attenuate 20 dB in low-frequency ranges to 45 dB in frequencies over 1000 Hz. Ear plugs can be less inexpensive than ear muffs, but they may not provide the same level of safety.

See Chap. 11 for detailed information on the hazards of high sound levels.

RESPIRATORS

Protection against inhalation hazards is often poor in laboratory safety programs. Respirators are sometimes used as inexpensive substitutes for effective ventilation, although good respiratory systems are not cheap.

For well-designed experiments in school science laboratories, routine respiratory protection should not be required. However, in case of accident or experiments where airborne toxic contaminants may exist, protective devices should be supplied.

Because the possibility of a serious accident always exists in any laboratory, some type of respiratory protection should be on hand at all times. Respiratory protection is furnished to the user or wearer by a mask that performs one or more of the following functions: filters particles, removes vapors and gases by adsorption, and supplies air or oxygen. The basic school respirator should be a dual cartridge device with a mechanical filtering capacity and a limited adsorption capacity for gases and vapors.

The primary method for the protection of laboratory personnel from airborne contaminants should be to minimize the amount of such materials entering the laboratory air. When effective engineering controls are not possible, suitable respiratory protection should be provided. It is the responsibility of the laboratory supervisor to determine when such protection is needed and to ensure that it is used.

Under OSHA regulations, only equipment listed and approved by the Mine Safety and Health Administration (MSHA) and the National Institute for Occupational Safety and Health (NIOSH) may be used for respiratory protection. Also under these regulations, each site on which respiratory protective equipment is used must implement a respirator program in compliance with the approved standard (29 CFR, 1910.134) (see also ANSI Z88.2).

Types of Respirators

Several types of nonemergency respirators are available for protection in atmospheres that are not immediately dangerous to life or health but could be detrimental after prolonged or repeated exposure. Other types of respirators are available for emergency or rescue work in atmospheres from which the wearer cannot escape without respiratory protection. In either case, additional protection may be required if the airborne contaminant is of a type that could be absorbed through or irritate the skin. For example, the possibility of eye or skin irritation may require the use of a full-body suit and a full-face mask rather than a half-face mask.

The general classes of respirator protective devices are particle-removing respirators, gas- and vapor-removing respirators, and atmosphere-supplying respirators. It is imperative that the proper respirator be selected; otherwise, no protection will result.

The choice of the appropriate respirator to use in a given situation will depend on the type of contaminant and its estimated or measured concentration, known exposure limits, and warning and hazardous properties (e.g., eye irritation or skin absorption). Once this information is available, an appropriate type of respirator can be selected by using a guide such as Table 10-2. The degree of protection afforded by the respirator varies with the type.

1. *Chemical cartridge respirators:* These respirators function by the entrapment of vapors and gases in a cartridge or canister containing a sorbent material. Some sorbents take out only specific substances and others remove whole classes of compounds. Activated charcoal is probably the most common adsorbent. These respirators must fit snugly to the face to be effective. Conditions that prevent a facepiece-to-face seal (e.g., temple pieces of glasses or facial hair) will permit contaminated air to bypass the filter, possibly creating a dangerous situation for the user. Tests for the proper fit of the respirator on the user should be conducted prior to its selection and verified before a worker or student wearing one enters the area of contamination.

   Chemical cartridge respirators can be used only for protection against particular individual (or classes of) vapors or gases as specified by the respirator manufacturer and cannot be used at concentrations of the contaminant above that specified on the cartridge. Also, these respirators cannot be used if the oxygen content of the

TABLE 10-2. Guide for Selection of Respirators
(Respirators are listed in order of decreasing protection efficiency.)

Type of Hazard	Type of Respirator
Oxygen deficiency	Self-contained breathing apparatus. Hose mask with blower. Combination of air-line respirator and auxiliary self-contained air supply or storage receiver with alarm.
Gas and vapor contaminants immediately dangerous to life or health	Self-contained breathing apparatus. Hose mask with blower. Air-purifying full-facepiece respirator chemical canister (gas mask). Self-rescue mouthpiece respirator (for escape only). Combination of air-line respirator and auxiliary self-contained air supply or storage receiver with alarm.
Gas and vapor contaminants not immediately dangerous to life or health	Air-line respirator. Hose mask with blower. Air-purifying half-mask or mouthpiece respirator with chemical cartridge.
Particulate contaminants immediately dangerous to life or health	Self-contained breathing apparatus. Hose mask with blower. Air-purifying full-facepiece respirator appropriate filter. Self-rescue mouthpiece respirator (for escape only). Combination of air-line respirator and auxiliary self-contained air supply or storage receiver with alarm.
Particulate contaminants not immediately dangerous to life or health	Air-purifying half-mask or mouthpiece respirator with filter pad or cartridge. Air-line respirator. Air-line abrasive-blasting respirator. Hose mask with blower.
Combination of gas, vapor, and particulate contaminants immediately dangerous to life or health	Self-contained breathing apparatus. Hose mask with blower. Air-purifying full-facepiece respirator chemical canister and appropriate filter (gas mask with filter). Self-rescue mouthpiece respirator (for escape only). Combination of air-line respirator and auxiliary self-contained air supply or storage receiver with alarm.
Combination of gas, vapor, and particulate contaminants not immediately dangerous to life or health.	Air-line respirator. Hose mask without blower. Air-purifying half-mask or mouthpiece respirator with chemical cartridge and appropriate filter.

Source: American National Standards Institute, Practices for Respiratory Protection, New York, 1969, Z88.2.

air is less than 19.5%, in atmospheres immediately dangerous to life, or for rescue or emergency work.

Because it is possible for significant breakthrough to occur at a fraction of the canister capacity, knowledge of the potential workplace exposure and length of time the respirator will be worn is important. It may be desirable to replace the cartridge after each use to ensure the maximum available exposure time for each new use. Difficulty in breathing or the detection of odors indicate plugged or exhausted filters or cartridges or concentrations of contaminants higher than the absorbing capability of the cartridge; such items should be replaced promptly (if necessary, by a more effective type of respirator).

Organic vapor cartridges must not be used for vapors that have poor warning properties or those that will generate high heats of reaction with the sorbent materials in the cartridge.

2. *Dust, fume, and mist respirators:* These are designed to protect the wearer against inhalation of material dispersed in the air as distinct particles such as dust, smoke of solid particulates, or a mist or fog of liquid droplets. Such particulate-removing respirators usually trap the particles in a filter composed of fibers. Air filters may be

simple cotton pads held by cassettes or complex folded mats held in canisters. The filter must be appropriate for the hazardous material in question. The respirator must fit the wearer's face and that person must be trained in its proper use.

These types of respirators are not 100% efficient in removing particles and can be used only for protection against particular individual or classes of particles as specified by the manufacturer. The use of any air-purifying respirator presupposes an adequate amount of oxygen in the air being breathed. Dust-removing respirators should be used only when no harmful gases exist.

Respirators of this type are generally disposable. Some examples are surgical masks and 3M[RT] toxic-dust and nuisance-dust masks, which can be used to filter out animal dander and nontoxic and nuisance dusts. Some are NIOSH-approved for more specific purposes such as protection against simple or benign dust and fibrogenic dusts and asbestos. The more expensive facepiece type offers the best protection against materials such as a nuisance dust like sawdust, a pneumoconiosis producer like silica, or a toxic like lead oxide. The useful life of the filter is dependent on the concentration of contaminant encountered.

Particulate-removing respirators such as surgical masks afford no protection against gases or vapors and should not be used when handling chemicals. They provide little if any protection and may give the user a false sense of security. They are also subject to the limitations of fit described above.

3. *Supplied-air respirators:* These supply fresh air or oxygen to the facepiece of the respirator at a pressure high enough to produce a pressure higher than the outside atmosphere. As a result, the supplied air flows outward from the mask and contaminated air from the work environment cannot readily enter the mask. This characteristic renders the face-to-facepiece fit less important than with other types of respirators. Fit testing is, however, required before selection and use.

   Atmosphere-supplying respirators are used when a lack of oxygen is anticipated or suspected and/or a high degree of protection is necessary. They are effective protection against a wide range of air contaminants (gases, vapors, and particulates) and can be used where oxygen-deficient atmospheres are present. They are free from the maintenance problems associated with charcoal, particulate, and chemical-scrubbing filters. Such respirators can be used where concentrations of air contaminants could be immediately dangerous to life or from which the wearer could not escape unharmed without the air of the respirator, provided (1) the protection factor of the respirator is not exceeded and (2) the provisions of 29 CFR 1910.134 (a safety harness and an escape system in case of compressor failure) are not violated.

   The breathing air is supplied with an "umbilical" tube. The air supply of this type of respirator must be free of contaminants (e.g., by the use of oil filters and carbon monoxide absorbers), and some consideration should be given to its quality and relative humidity. Most laboratory air is not suitable for use with these units. The umbilical lines are long lengths of hose connected to the air supply. The user must drag this line around and the range of their use is limited to the maximum length of hose specified by the manufacturer.

4. *Self-contained breathing apparatus:* This is the only type of respiratory protective equipment suitable for emergency or rescue work. This equipment consists of a full-face mask connected to a cylinder of compressed air and has no limitations with regard to its use in areas of toxic contaminants or oxygen deficiency. However, the air supply is limited to the capacity of the cylinder (5–30 min use time) and, therefore, these respirators cannot be used for extended periods without recharging or replacing the cylinders. For safety reasons, the "pressure/demand" type, which always has a positive pressure within the mask, is much preferred to the "demand" type. Also, they are bulky and heavy, and additional protective apparel may still be required depending on the nature of the hazard. All institutions or organizations with laboratories in which chemicals are used should have protective equipment of this type available for emergencies and provide training in its use to selected personnel.

Procedures and Training

Each area where respirators are used should have written information available that shows the limitations, fitting methods, and inspection and cleaning procedures for each type of respirator available. Personnel who may have occasion to use respirators in their work must be thoroughly trained in the fit testing, use, limitations, and care of such equipment. Training should include demonstrations and practice in wearing, adjusting, and properly fitting the equipment. Contact lenses should not be worn when a respirator is used, especially in a highly contaminated area. OSHA regulations require that a worker be examined by a physician before beginning work in an area where a respirator must be worn [29 CFR, 1910.134(b) (10)].

Respirators for routine use should be inspected before each use by the user and periodically by the laboratory supervisor. Self-contained breathing apparatus should be inspected once a month and cleaned after each use. Defective units should not be used but should be repaired by a qualified person or replaced promptly.

STORAGE AND INSPECTION OF EMERGENCY EQUIPMENT

It is often useful to establish a central location for the storage of emergency equipment. Such a location should contain the following:

1. Self-contained breathing apparatus
2. Safety belt with rope (to maintain contact with rescuers entering a laboratory under emergency conditions)
3. Blankets for covering injured persons
4. Stretchers
5. First-aid equipment (for unusual situations such as exposure to cyanide, where immediate first aid is required)

Safety equipment should be inspected regularly (e.g., every 3 to 6 months) to ensure that it will function properly when needed. It is the responsibility of the laboratory supervisor or safety coordinator to establish a routine inspection system and to verify that inspection records are kept.

1. Fire extinguishers should be inspected for broken seals, damage, and low gauge pressure (depending on the type of extinguisher). Proper mounting of the extinguisher and its ready accessibility should also be checked. Some types of extinguishers must be weighed annually, and periodic hydrostatic testing may be required.
2. Self-contained breathing apparatus should be checked at least once a month (and after each use) to determine whether proper air pressure is being maintained. The examiner should look for signs of deterioration or the wear of rubber parts, harness, and hardware and make certain that the apparatus is clean and free of visible contamination.
3. Safety showers and eyewash fountains should be examined visually and their mechanical function tested.

RECORDKEEPING

Science teachers often keep some type of written record of the personal protective equipment that is issued to students. These records are kept to control equipment losses rather than provide maintenance information. Inventory records on personal protective equipment should be maintained, however, because many types of personal protective equipment require preventative maintenance. For instance, protective eyewear requires regular cleaning and disinfection.

If proper maintenance procedures are not performed on the personal protective equipment requiring them, the equipment may actually magnify a health hazard or serve as the source of a new hazard. Respirators with expired cartridges can create a false sense of security in the wearer. The wearer is seriously exposed to chemical hazards while thinking he or she is protected. Dirty respirators and safety glasses can serve as the source of infectious microorganisms or skin-penetrating, toxic chemicals. Records that are kept up-to-date will enable the instructors and students to make informed decisions regarding the condition and suitability of equipment for use. The compilation of this type of written record is an integral part of a safety and health program. Students may be effectively integrated into this portion of the program. With supervision, they can be given the responsibility for keeping the written inventory record and for performing the required maintenance procedures on the equipment.

There is no recommended format for keeping records of this type. A suitable record form would be easy to construct. It can be made a student safety project.

SUGGESTED READINGS

American National Standards Institute, Z87.1, American National Standard for Occupational and Educational Eye and Face Protection, New York. Use the latest standard.

American National Standards Institute, Z88.2, American National Standard on Practices for Respiratory Protection, New York. Use the latest standard.

National Society to Prevent Blindness, Model School Eye Safety Law, Schaumburg, Ill.

Title 29 of the Code of Federal Regulations, Part 1910.134, Respiratory Protection.

CFR titles (chapters) are available from the U.S. Government Printing Office, Washington, D.C. Each title is revised annually.

11

LIGHT AND OTHER FORMS OF RADIATION

THE ELECTROMAGNETIC SPECTRUM

There are several kinds of radiation propagated by photons that are included in the electromagnetic spectrum. Radiations from the portion of this spectrum that are of importance and interest to laboratory workers are visible light, ultraviolet and infrared radiations, microwaves, and ionizing radiations (which have wavelengths shorter than ultraviolet light). There is only one narrow band of these waves in the entire spectrum that can be seen as visible light. All the rest are invisible. Radiations from electromagnetic sources differ in frequency, wavelength, energy level, their visual response by the human eye, and their effects on the human body. Exposures to some of the radiations may be harmful to man, depending on the intensity and duration of the exposure. On the other hand, certain curative radiations can be derived from various electromagnetic sources.

Wavelength is the one physical factor of these radiations that differentiates one from another. The units of length adopted by physicists for measuring these extremely small wavelengths are the nanometer (nm) or the Angstrom (Å) unit. Centimeters (cm) and Hertz (Hz) are also used as units of measurement for these wavelengths. Figure 11-1 gives the types of radiation and the various ranges of wavelengths in nm and Hz. One nanometer equals 10 Angstrom units.

The following definitions are customary:

X- and γ-rays, shorter than 10 nm (1 nm = 10^{-9} m)

ultraviolet, 10–400 nm

visible, 400–700 nm

infrared, 700–1,000,000 nm (1 mm)

microwaves, 1 mm to 1 m

radio waves, 1–10,000 m

electric waves, longer wavelengths

NONIONIZING RADIATION

Nonionizing radiation is electromagnetic radiation with wavelengths longer than 10 nm (ultraviolet and longer). Photons of this energy range do not have enough energy to remove an electron from most materials when they are absorbed. However, some of these photons do have enough energy to cause other effects that can be harmful.

Ultraviolet Radiation

Ultraviolet (UV) radiation is an invisible radiant energy that is produced by natural and artificial sources and accompanies much visible light. Ultraviolet radiation is divided into subregions:

Ultraviolet Region, 10–400 nm

	Subregion	Approximate Wavelength Range (nm)
Near	UVA	315–400
Mid	UVB	280–315
Far	UVC	100–280
	Ionization	10–100

Other descriptive terminology in use separates ultraviolet sources into:

1. Long wave, which is continuous to the visible light spectrum and extends from 290–390 nm.
2. Short wave, which extends from 180–290 nm.

The UVA region or that portion of the long waves that ranges from 310–410 nm is also known as the black light portion of the ultraviolet region, although this is not a very specific term. The peak transmission of a black light lamp occurs around the 365-nm mercury line.

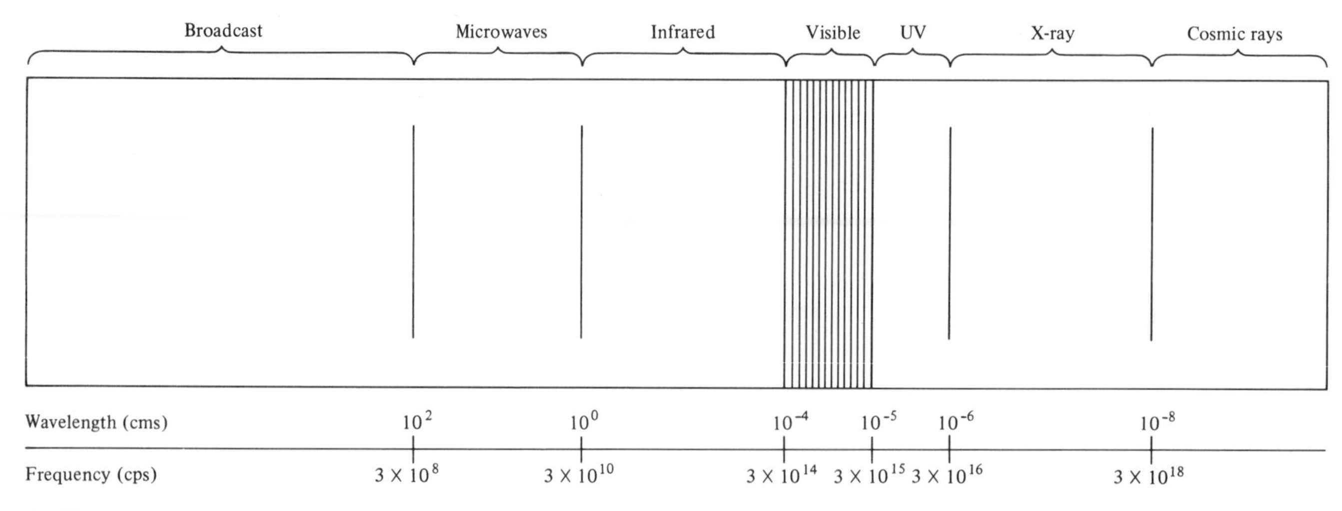

FIGURE 11-1. Electromagnetic spectrum. (*Source:* NIOSH, *Safety in the School Science Laboratory, Instructor's Resource Guide,* Cincinnati, Ohio, 1979, p. IV–15.)

Since the wavelengths below 190 nm are strongly absorbed by air (in molecular excitation transitions), they are not easily observed except in vacuum, so this range is called the vacuum ultraviolet. Ozone (O_3) is formed from molecular oxygen (O_2) by excitation and collision combination at wavelengths from 170–230 nm. The actinic and keratitic regions are so-called because they produce biological effects on the skin. The skin erythemal effects are produced mainly by mid-ultraviolet photons 250–315 nm.

Very long ultraviolet wavelengths from 310–320 nm penetrate deeper into the skin but exert little biological effect. Short rays below 230 nm exert very little biological effect and have practically no penetration. Most biological action occurs in the intervening zone between the long wavelengths and the short rays.

Physiological Effects. Biological effects of ultraviolet radiation include tissue damage to the eyes and skin. A typical example of the injurious effects produced by ultraviolet radiation is sunburn, which results from the naturally occurring ultraviolet rays produced by the sun. How serious a sunburn is depends on the length of the exposure and intensity of the radiation, as well as an individual's sensitivity. Continual exposure to ultraviolet radiation speeds skin aging and can cause skin cancer. Prolonged exposure to ultraviolet energy from 280–340 nm, with the greater potency from 280–300 nm, is capable of causing skin cancer.

Exposure to the eyes is particularly dangerous because the ultraviolet radiation cannot be seen or, at first, felt. Consequently, an exposed individual is not always aware that his or her eyes are being affected.

Radiations below 295 nm are absorbed in the cornea and conjunctiva. The cornea is most sensitive to ultraviolet at 270 nm. Overexposure of the eye to this ultraviolet radiation produces an inflammatory condition in the cornea referred to as photokeratitis. This condition usually follows a characteristic course. As with skin erythema (sunburn), there is a latent post-irradiation period, typically 4–8 hr, before the appearance of acute clinical symptoms. The symptoms include conjunctivitis accompanied by an erythema of the skin of the face and eyelids. The conjunctivitis is extremely painful and, although usually temporary, can cause permanent injury to the eyes. There is the sensation of "sand" in the eyes, photophobia, lacrimation, and blepharospasm (involuntary closure of the eyelids). Secondary infection may result if the individual affected attempts to allay the painful irritation by rubbing his or her eyes. Corneal ulcers and inflammation of the iris can also result.

Ultraviolet radiation can cause the conversion of atmospheric oxygen to ozone, which is an extremely strong oxidizer and thus can be an air pollutant. Proper ventilation should be provided to remove ozone and other gases that may be created by ultraviolet-generating equipment.

Sources. The sun is the major natural source of ultraviolet radiation but many artificial sources are used in industry, medicine, and research. Schools and laboratories use a variety of ultraviolet-producing equipment, including arc welders, welding and cutting torches, arc lamps (including carbon arcs), spectrograph sources and other analytical devices, mercury vapor lamps, germicidal lamps, xenon lamps, deuterium lamps, chemical synthesis equipment, furnaces, and some photocopying machines. All these sources can produce enough light to damage the eye.

One of the most prevalent sources of exposure to ultraviolet radiation occurs during welding operations. In particular, inert-shielded gas welding produces much higher levels of ultraviolet than other types of welding. Arc welding produces ultraviolet radiation in broad bands, which often appear as a continuous spectrum. The intensities of the various bands depend on many factors: materials used in the electrodes, discharge current, and gases surrounding the arc.

High-pressure mercury vapor lamps are used in photo-chemical reactions, mineral identification, to produce fluorescence, and for the diagnosis of dermal and scalp disorders, including porphyria (disturbance of porphyrin metabolism). Quartz-mercury arcs emit radiation over much of the ultraviolet spectrum and can cause erythema and conjunctivitis from radiation over the range of 200–320 nm.

High-pressure xenon arcs emit a spectrum like that of sunlight. Carbon arcs emit a continuous spectrum from the incandescent electrodes, upon which a broadband spectrum from the luminous gases is superimposed.

The plasma torch can produce temperatures over 6000 K, the temperature at the surface of the sun, and intense ultraviolet radiation can result. Exposure to radiation from plasma torches can result in keratoconjunctivitis and sunburn if the skin and eyes are not protected.

Many laboratory analytical instruments use ultraviolet lamps for measurements (e.g., spectrometers and fluorometers).

Instruments for Measuring Ultraviolet. The measurement of ultraviolet light is considerably more complicated than the measurement of visible light, since different wavelengths of ultraviolet have different effects. For instance, 253.7 nm is the most effective bactericidal wavelength, whereas 294.7 nm has the greatest epidermal effect. Therefore, in the measurement of ultraviolet, it is important to determine both the intensity and wavelength of the radiation.

General Control Measures. Anyone operating equipment that produces ultraviolet radiation should wear protective clothing, gloves, and face shields. Enclosures or shields opaque to ultraviolet radiation can also be used. During welding operations that generate ultraviolet light, side and back screens should be used to protect nearby people. Warning signs should be placed to alert people in areas where there are open arcs or other high-intensity light-emitting sources.

Eye Protection. Eye protection filters for glass workers, steel and foundry workers, and welders were developed empirically; however, optical transmission characteristics are now standardized as shade numbers and specified for particular applications. Although maximum transmittances for ultraviolet and infrared radiation are specified for each shade, the visual transmittance or visual optical density OD defines the shade number S by the relation in eq. (1).

$$S = 7/3\ OD + 1 \tag{1}$$

where $OD = \log_{10} (E_0/E)$
E_0 = intensity of the incident beam
E = intensity of the transmitted beam

Table 11-1 is a guide to the selection of eye protection filters.

Infrared Radiation

Infrared radiation consists of a portion of the electromagnetic spectrum with wavelengths longer than that of visible light. Infrared radiation is divided into subregions:

Infrared Region, 700 nm–1,000,000 nm

	Subregion	Approximate Wavelength Range (nm)
Near	IRA	700–1,400
	IRB	1,400–3,000
Far	IRC	300–1,000,000

The cornea of the eye is highly transparent to energies between 750 nm and 1300 nm and becomes opaque to radiant energy above 2000 nm. Thermal damage to the cornea is dependent on the absorbed dose and probably occurs in the thin epithelium.

An iris dose of about 4.2 W·sec/cm^2 has been suggested as the minimum producing damage from a source emitting principally between 800–1100 nm, which would require a dose to the corneal surface of 10.8 W·sec/cm^2 for cataract formation. However, corneal damage has been reported at about 7.6 W·sec/cm^2 in the range 800–1100 nm. Damage to both the cornea and iris is an acute condition, and there are no data available that indicate chronic effects in these parts of the eye. The lens is heated by the heated iris and is unable to keep cool by blood flow at this level of energy absorption. The local heating of the lens is thought to

TABLE 11-1. Guide for the Selection of Proper Shade Numbers During Welding

Welding Operation	Shade No.
Shielded metal-arc welding:	
1/16, 3/32, 1/8, 5/32 in. electrodes	10
Gas-shielded arc welding (nonferrous):	
1/16, 3/32, 1/8, 5/32 in. electrodes	11
Gas-shielded arc welding (ferrous):	
1/16, 3/32, 1/8, 5/32 in. electrodes	12
Shielded metal-arc welding:	
3/16, 7/32, 1/4 in. electrodes	12
5/16, 3/8 in. electrodes	14
Atomic hydrogen welding	10–14
Carbon arc welding	14
Soldering	2
Torch brazing	3 or 4
Light cutting, up to 1 in.	3 or 4
Medium cutting, 1–6 in.	4 or 5
Heavy cutting, 6 in. and over	5 or 6
Gas welding (light), up to 1/8 in.	4 or 5
Gas welding (medium), 1/8 to 1/2 in.	5 or 6
Gas welding (heavy), 1/2 in. and over	6 or 8

Note: In gas welding or oxygen cutting where the torch produces an intense yellow light, it is desirable to use a filter or lens that absorbs the yellow or sodium line in the visible light of the operation.
Source: From NIOSH, *Nonionizing Radiation 583,* 1977, p. 16–63.

produce cataracts. For the sake of convenience, a cataract can be defined as an opacity of the crystalline lens or its capsule, which may be developmental or degenerative, obstructing the passage of light. The degenerative cataract is a manifestation of aging, systemic disease, trauma, or certain forms of radiant energies among other causes.

A common source of high-intensity infrared radiation is the infrared laser. The section on lasers contains more information on infrared radiation hazards. See the "Ultraviolet Radiation" section for information on eye protection filters.

Microwave and Radio Frequency Radiation

The term microwave or radio frequency (RF) radiation is generally considered to apply to that portion of the electromagnetic spectrum ranging from 100–30,000 megahertz (MHz) or wavelengths from 3m to 1 cm. This wide range of frequencies includes commercial as well as television and broadcast bands; bands of diathermy and microthermy units; and the X (9000–9500 MHz), S (2600–3200 MHz), and L (1100–1400 MHz) bands of radar.

Microwave radiation is sometimes confused with ionizing radiation. This is unfortunate since the two radiations have no important similarities as far as biological effects are concerned. Microwaves have some of the characteristics of infrared radiation in that they produce localized heating of the skin; however, they penetrate deeper than infrared radiation. In general, the heating produced is proportional to the field intensity of this radiation. Other factors influencing the effects of microwave radiation include:

1. Frequency or wavelength of the radiation
2. Period or exposure time
3. Air currents and environment temperature
4. Body weight, type, or mass in relation to the exposed area
5. The irradiation cycle rate, which refers to the individual ON/OFF periods during a unit time interval (1 min) when the total time of irradiation per minute is kept constant
6. Orientation or position of an individual influencing resonant conditions and standing waves
7. Difference in sensitivity of organs and tissues
8. Effect of reflections

Biological Effects. Although the exact nature of the biological effects of microwave radiation is not completely understood, most of the experimental data support the concept that the effects are primarily due to local or general hyperthermia, that is, heating. Some nonthermal physiological effects may also occur.

Biological damage is related to wavelength and power intensity. In general, the longer wavelengths (10 cm or more) are more deeply penetrating than the short wavelengths (less than 10 cm). The longer wavelengths produce a greater temperature rise in a significant volume of tissue than the shorter wavelengths, but the heating is less perceptible for a given power intensity.

The known biological effects of microwave radiation include:

1. Whole-body heating
2. Cataract formation (damage to the lens of the eye)
3. Testicular damage

Of these three, cataract formation is the area of greatest concern and the lens of the eye is the critical organ.

Thermal Effects. The biological effects of microwave radiation are mostly the result of the conversion of electromagnetic energy into heat energy within the tissue itself. The effects are the same as with any burn, but the heat may be generated deep in the tissue rather than on the surface. Since temperature is related directly to the rate and extent of molecular vibrations, a large part of the increase in temperature is due to the increased vibration of polar molecules such as water, which is the major constituent of biological systems. Heat generation (and thus microwave absorption) is greatest in those tissues with a high content of water and is enhanced locally in the areas adjacent to bone or tough muscle planes that act as reflecting surfaces. Such local reflection and possible superposition effects can produce localized heating. Thermal effects are unlikely at intensities below 1 mW/cm^2.

The depth of penetration is related to the wavelength of radiation and also differences in the thermal conductance of various tissue. These result in different amounts of energy being absorbed and different degrees of heating of such tissue as fat, bone, and muscle. This is a dynamic situation, and as the energy is being absorbed and converted into heat, the heat is dissipated by the blood stream. Thus, the heating effect is most likely to cause damage to organs where normal heat control is poor and blood circulation is low. The organs most likely affected are the eyes and testes.

Within the critical wavelengths, the potential hazard to the eye is one of the most serious aspects of microwave exposure. In addition to the relatively poor blood supply, there are cavities near the eye that contain intraocular fluids with high electrical conductivity. Therefore, there is short penetration, or high absorption, of electromagnetic radiation in the eyeball. The lens, being avascular and enclosed in a capsule, is at a disadvantage by not having a cooling system (blood flow) as do other tissues. Not having mechanisms to remove dead cells or to replace cells, the lens cannot repair itself as do other tissues in the body. Thus, damage to the lens is generally irreversible. Depending on the dose, the damaged cells slowly lose their transparency and opacity (cataract) may not occur until sometime after exposure. Frequencies of 3000 MHz are the most efficient at heating the lens of the eye. U.S. researchers generally would agree that for a single exposure of less than 1 hr, with 48 hr before another such exposure, the threshold below which damage does not occur is about 80 mW/cm^2. Other researchers believe that there are cumulative effects of lower-power density exposures over long periods of time.

Both direct and indirect effects of heating occur. Direct heating of tissues is a primary effect, whereas an effect caused by a systemic reaction to that heating would be secondary, or indirect. A general heat stress due to whole body heating or damage to some specific body tissue due to local heating are examples of primary effects. Adaptation stresses and endocrine system or central nervous system (CNS) reactions to the primary heating are examples of secondary effects.

Nonthermal Effects. While certainly the most readily observed effects of the absorption of microwaves in biological systems are thermal, there is also some evidence of nonthermal effects. Several investigators have pointed out that microwaves may also interact with biological material without the production of a significant amount of heat. According to Soviet researchers, the exposure of animals to microwaves of low intensity, which do not produce any appreciable thermal effect, can lead to functional changes mainly in the nervous and cardiovascular systems. Other reported nonthermal biological interactions include "pearl chain" formations of blood cells, mutations in exposed garlic root tips, changes in proteins in human gamma globulin, sounds "heard" by several people that correspond to the frequency of modulation of an incident microwave beam and occasionally stomach distress in humans. Generally, these data are uncertain in nature, quality, and repeatability. Some of these effects may be explained by localized heating, or secondary effects such as adaptation stresses. Many researchers are studying this problem in laboratories around the world.

Other Effects. Metal-containing surgical implants can also be a problem. Electromagnetic (EM) energy is readily absorbed by conductive materials. Thus, an electrical conductor that is part of an artificial organ or pacemaker implanted in the body will absorb this energy, which induces electrical currents in the conductor. Pacemakers are sometimes implanted in people with heart trouble to assist their hearts in maintaining a pumping rhythm. Pulsed microwave energy at even low power levels can be hazardous to people with pacemakers, since the spurious currents induced may cause the heart to go out of synchronization. Older pacemakers generally present the greatest problem. More recent designs are less susceptible. However, no levels of microwave power can be considered safe for people with pacemakers. They should be excluded from areas around unshielded pulsed microwave sources.

Metal pins or other orthopedic implants may heat up in high electromagnetic fields and cause pain or injury. Physical damage is more likely to result if the local blood flow is not sufficient to dissipate the heat.

Microwave Ovens. Microwave ovens operate at either 915 or 2450 MHz. Most manufacturers use the 2450 MHz band for these ovens. The oven typically has a magnetron source, a means of coupling radio frequency energy, controls for turning off and on at desired times, a stirrer to produce a uniform field inside the oven, and a gasket seal to prevent radio frequency field leakage. Also, there is an interlock switch to turn off the power automatically if the door is opened.

In surveying a microwave oven for radiation leakage, one uses a standard probe (antenna) and checks the various possible apertures for microwave leakage, for example, the door seals, vents, etc.

Maximal Safe Exposure Value. In the United States, 10 mW/cm^2 is accepted for continuous daily exposure, with no exposure allowed above 100 mW/cm^2. In the Soviet Union, 0.01 mW/cm^2 is accepted for continuous exposure, 0.1 mW/cm^2 is permitted for 2 hr, and 1 mW/cm^2 for no more than 20 min. These numbers do not take into consideration the following:

1. The duration of the exposure or repeated short-term exposure through any given period. Exposure time is of special concern in the case of pulsed transmissions of radar.
2. The frequency or wavelength of the radiation, even though experimental data have indicated that the coefficient of absorption in tissue varies with frequency.
3. Exposure to radiation of several different frequencies. This is of prime importance at radar installations where microwave generators operate at different frequencies and could be a source of potential multifrequency exposure.
4. Increased dissipation of the heat caused by air movement past the body being exposed to the radiation.
5. The differences in sensitivity of various organs to the absorption of heat.
6. Consideration of reflected radiation.
7. Environmental temperature and its effect on the dissipation of heat created by the impingement of microwave radiation on the body.

Control Methods. Several specific methods have been developed and applied to control potential exposure to microwave radiation and should be given careful consideration:

1. To increase the distance between the source of energy and the workers. Microwave radiation normally can be considered as a point source and the power density at any given location can be estimated using the inverse square law. Care, however, must be exercised in the use of the inverse square law as a control measure near a microwave transmitter, since the generator does not act as a point source in the near field and crossover regions near the antenna or reflector. The reflection of radiation from objects can increase and intensify the energy ten to twelve times the emitting power of the generator.
2. Shielding of occupied areas by the use of wire mesh for reflecting microwave radiation, which has proved to be a satisfactory method of control (see Table 11-2). Shielding of this type should be used primarily as a means of control in the far field and preferably in areas of occasional occupancy. Another form of shielding is to blank off certain sectors of the microwave field to assure that personnel will not be exposed while in the sector. This type of shielding may reduce the effective operational capability of microwave generators since the area traversed has been reduced. This type of shielding possibly could be eliminated by the proper positioning of the transmitters.
3. The use of protective clothing by operational and maintenance personnel. Proper protective clothing has been developed for workers in fields of high power densities. This protective clothing consists of a suit made of a special metalized fabric that is coated with neoprene to protect against the discharge of electrical currents. Special closures are provided since metalized zippers cannot be used. Mittens and boots are fitted to the suit to make it a completely contained unit. The eyes are protected by closely woven copper screening. It is difficult to use a suit of this type, particularly in warm or high temperatures, unless a ventilating or air-conditioning system is provided. Protective clothing as a means of control for microwave radiation is somewhat open to question and its use as a substitute for other engineering controls and safe operating procedures should never be condoned.

TABLE 11-2. Microwave Shielding Properties at 2450 mHz

Material	Attenuation Factor
60 × 60 Mesh, metal screening	0.997
32 × 32 Mesh, metal screening (window type)	0.994
8 in. Solid concrete block	0.994
Window glass	0.37
3/8 in. Plywood	<0.37

Source: From NIOSH, *Nonionizing Radiation 583,* 1977, p. 14–99.

A warning sign should be placed outside any area where a potentially harmful level of microwave radiation may be encountered.

LASERS

The term laser is an acronym derived from Light Amplification by Stimulated Emission of Radiation. The effects of laser radiation are essentially the same as light generated by more conventional ultraviolet, infrared, and visible light sources. The unique biological implications attributed to laser radiation are generally those resulting from the very high intensities and monochromaticity of laser light. Such sources differ from conventional light emitters primarily in their ability to attain highly coherent light (in phase). The increased directional intensity of the light generated by a laser results in concentrated light beam intensities at considerable distances.

Biological Effects

Laser radiation impinging on the body is reflected, absorbed, and transmitted; the percentage of each depends on the characteristics of each particular kind of tissue at the wavelengths of concern. The absorption of laser radiation by living tissue can result in temporary or permanent damage to the tissue, ranging from mild erythema to blistering and/or charring. The effects depend on the total energy absorbed and the rate at which it is absorbed. In general, the extent and significance of injury depend on the site irradiated, characteristics of the tissue, and characteristics of the incident radiation. The tissue characteristic of primary importance is its absorption coefficient, that is, ability to absorb radiation. Radiation characteristics of primary importance are wavelength, exposure duration, pulse width, repetition rate, beam intensity, and degree of beam focusing.

Tissue damage caused by absorption of laser radiation of low irradiance levels and in the visible and infrared spectral region appears to be a result of increased tissue temperature. However, in the ultraviolet spectral region, damage appears to result primarily from photochemical reactions, and injury from coherent laser radiation is similar to that produced by noncoherent sources of electromagnetic radiation of comparable wavelengths. (See the section on ultraviolet radiation.) High levels of irradiance may cause damage due to other effects such as shock waves or high electrical field strengths. Unnecessary exposure to laser radiation should be avoided, regardless of the level of irradiance. The tissues of the eye and skin are most susceptible to laser radiation.

Eye Effects. As the eye is often far more vulnerable to injury than the skin from visible and near-infrared laser radiation, it is considered the organ most important to protect from all wavelengths of laser radiation. Because the eye focuses radiation in the visible and near visible region of the spectrum, the irradiance at the retina may be as much as 100,000 times greater than at the cornea. Such a concentration of radiant power can cause the retina to be burned in much the same way that a piece of paper can be set ablaze when a magnifying glass focuses the rays of the sun. As a result, injury to the retina can occur, in the visible region, at radiation levels that are far less than those which produce corneal or skin damage. Retinal damage may vary from minimally detectable to severe, with permanent loss of vision possible.

Generally three types of eye damage may result from exposure to wavelengths between 400 and 1400 nm (visible and near infrared): structural, thermal, and photochemical. The type of damage depends on the wavelength, power, and exposure time. Structural and thermal damage depend on high power levels and short exposure times. They are independent of wavelength. Thermal damage does not occur without a significant rise in temperature (approximately 10°C). Photochemical damage occurs even when the retinal temperature increase is less than 0.1°C. On the other hand, radiation from the ultraviolet and infrared portions of the spectrum is more likely to produce damage to the cornea and lens.

Skin Effects. Superficial tissue damage can result from the production of heat by the absorption of radiation on the skin. This is usually considered to be primarily a surface phenomenon. The pigmentation of the tissue affects the magnitude of biological reaction with a greater reaction to areas of hyperpigmentation. This reaction has been observed from focused and unfocused laser beams.

Control Methods

The fundamental objective of controlling laser hazards is to limit the possibility of a potentially hazardous exposure, particularly to unaware transient personnel, and to provide reasonable and adequate guidance for the safe use of lasers and laser systems.

The methods of engineering control of exposure to lasers are based on the same principles as are applied to other electromagnetic radiation. It must be remembered, however, that coherent light does not follow the inverse square law. Thus, the use of distance from the source is not as effective a means of control as it is with other electromagnetic radiation. The necessity for engineering control is based on the intensity of the energy and the duration of exposure.

In establishing laser control measures, the following factors determine the type and amount of control necessary:

1. Power or energy output of the laser
2. Pulse length
3. Pulse repetition rate
4. Wavelength
5. Beam path
6. Beam shape (divergence, hot spots, atmospheric effects)
7. Number of laser systems at a particular location
8. Position of windows, doors, laboratory layout
9. Degree of isolation of location
10. Type of population (informed staff, students, uninformed transients)

In addition to the above factors, control measures also depend on laser classification (see Table 11-3).

Various techniques for the engineering control of potential hazardous laser sources have been developed and should be given careful consideration. These are:

1. Avoid exposure to the primary beam and any reflected energy from walls or other surfaces. This is a basic means of control; however, if this is not possible due to operational procedures, consideration should be given to other means of control. It must be remembered that an acceptable maximum exposure limit has not been determined and that one exposure may be sufficient to cause retinal damage.

2. Visual or audible signals should be used to indicate when the laser is operating. These signals should be positioned to call attention away from the light source and not toward the source of energy. If the laser is to be used outside of a building, the proposed path of the beam should be well posted and other special precautions should be required to reduce the possibility of exposure, as damage may occur at extended distances from some generators of coherent energy.

3. Laser and operations of high intensity which occur in a room may dictate that the walls and other surfaces be covered, if possible, with suitable material to reduce the reflection of the energy. In addition, the room should be well lighted to reduce the size of the pupil of the eye.

4. It may be desirable to enclose the operation in a suitable shield to reduce the possibility of expo-

TABLE 11-3. Classification of Laser Equipment

Class	Hazard	Description
1	Inherently safe	Exempt lasers. Considered incapable of producing damaging radiation levels and therefore exempt from any control measures or other forms of surveillance. Maximum permissible exposure levels cannot be exceeded.
2	Low-risk	Low-power lasers. Hazardous if looked at continuously. May be viewed directly, but must have a caution against continuous intrabeam viewing. Emission limited to 1 mW for less than 0.25 sec between 400 and 700 nm; hazards are prevented by aversion reflexes.
3A	Low-risk	Limit up to five times that for Class 2; viewing by the unaided eye is safe, but the use of optical instruments may be hazardous. Requires control measures that prevent viewing of the direct beam.
3B	Medium-risk	Emission limit higher still; any direct viewing may be hazardous, but not viewing by diffuse reflection. Requires control measures that prevent viewing of the direct beam.
4	High-risk	High-powered systems. Emission limit higher still; viewing by diffuse reflection may be hazardous. Skin injuries and fire hazard are also possible. Requires the use of controls that prevent exposure of the eye and skin to the direct and diffusely reflected beam.

sure to either the operational staff or transient personnel.

5. The use of protective eye shields is recommended, but the proper lens must be selected for the wavelength of the coherent energy being produced. Care must be taken in the use of eye protection filters as the transmittance characteristics of the material may change with repeated use or age.
6. One control of prime consideration is the education of workers regarding the health problems related to the use of coherent light and operational procedures that they must use to assure that they are not exposed excessively to this type of energy. Furthermore, operating personnel should be aware of the necessity of following such industrial medical procedures as reporting immediately any exposure to the medical department. This may be of particular importance if an after-image of the light source persists.

Specific Control Measures. To reduce the controls required and the potential hazard from a laser source, a complete enclosure of the laser beam (an enclosed laser) should be used when feasible. A closed installation (any location where lasers are used that is closed to transient personnel during laser operation) provides the next most desirable hazard control measure. Specific control measures to reduce the possibility of exposure of the eye and skin to hazardous laser radiation and to other hazards associated with the operation of those devices are outlined in the American National Standard for the Safe Use of Lasers (ANSI Z-136.1, 1973), some of which are excerpted below:

Class I: Exempt Lasers and Laser Systems. No control measures or warning labels are required; however, any needless direct exposure of the eye should be avoided as a matter of good practice. These lasers have such low power that if all the output were focused on the retina for a whole day, there would be no adverse effect.

Class II: Low-Power Visible Lasers and Laser Systems. These lasers must have a light to indicate operation and an appropriate warning label affixed to a conspicuous place on the laser housing or control panel, or on both the laser housing and control panel. These low-power lasers must be visible (400–700 nm) and are incapable of causing eye injury within the duration of the blink, or average response (0.25 sec). A hazard only exists if an individual overcomes his or her natural aversion to bright light and stares directly into the laser beam. The majority of low-power lasers today are helium neon devices (He-Ne) with a continuous wave power of 1 mW or less. The two operating safety rules are the following:

1. Do not permit a person to stare at the laser from within the beam.
2. Do not point the laser at a person's eye.

Class III: Medium-Power Lasers and Laser Systems. These lasers present serious potential for eye injury when the beam is viewed directly. They usually are not a hazard when the diffuse reflection is viewed, when the skin is exposed, nor do they pose a fire hazard. A class III system must have engineering controls and appropriate safety mechanisms as an integral part of the system. Examples include beam stops, beam enlarging systems, enclosures, shutters, interlocks, and so forth. Control measures focus on removing the possibility of viewing the beam directly. The following rules, excerpted from Chang, 1986, are useful:

1. Never aim a laser beam at a person's eyes.
2. Use proper safety eyewear if there is a chance that the beam or a hazardous specular reflection will expose the eyes.
3. Permit only experienced personnel to operate the laser. Do not leave an operable laser unattended if there is a chance that an unauthorized person may attempt to use it. A key switch should be used. A warning light or buzzer should indicate when the laser is operating.
4. Enclose as much of the beam path as possible.
5. Avoid placing the unprotected eye along or near the beam axis. The chance of hazardous specular reflections is greatest in this area.
6. Terminate the primary and secondary beams, if possible, at the end of their useful paths.
7. Use beam shutters and laser output filters to reduce the beam power to less hazardous levels when the full output power is not required.
8. Assure that any spectators are not potentially exposed to a hazardous condition.
9. Attempt to keep laser beam paths above or well below either sitting or standing position eye level.
10. Do not permit tracking of nontarget vehicles or aircraft if the laser is used outdoors.
11. Label lasers with appropriate Class III danger statements and placard hazardous areas with danger signs if personnel can be exposed.

12. Mount the laser on a firm support to assure that the beam travels along the intended path.
13. Assure that individuals do not look directly into a laser beam with optical instruments unless an adequate protective filter is present within the optical train.
14. Eliminate unnecessary specular (mirrorlike) surfaces from the vicinity of the laser beam path, or avoid aiming at such surfaces.

Class III, IV, and V Lasers and Laser Systems. These lasers present the most serious of all laser hazards. Normally, only well-enclosed high-power lasers are found outside of the research laboratory. Besides presenting a serious eye and skin hazard, these lasers can often ignite flammable targets, create hazardous airborne contaminants, and usually have a potentially lethal high-current, high-voltage power supply. Try to enclose the whole length of the laser beam. If this is done, the laser device could revert to a less hazardous classification. For high-power lasers that cannot be completely enclosed, these rules should be carefully followed:

1. Access to the laser room must be strictly controlled.
2. The laser beam should be terminated by a beam stop at the end of the useful path.
3. Indoor operations should be confined to a light-tight room with entrances interlocked to the laser so that the beam is blocked when the door is open.
4. Insure that all personnel wear adequate eye protection, and if the laser beam irradiance represents a serious skin or fire hazard, that a suitable shield is present between the laser beam(s) and personnel.
5. Use remote firing and video monitoring or remote viewing through a laser safety shield when feasible.
6. Use beam traverse and elevation stops on outdoor laser devices such as LIDAR (Light Detection and Ranging) to assure that the beam cannot intercept occupied areas or intercept aircraft.
7. Use beam shutters and laser output filters to reduce the laser beam irradiance to less hazardous levels whenever the full beam power is not required.
8. Assure that the laser device has a key-switch master interlock to permit only authorized personnel to operate the laser.
9. Install appropriate signs and labels.
10. Remember that optical pump systems may be hazardous to view and that once optical pumping systems for pulsed lasers are charged, they can spontaneously discharge, causing the laser to fire unexpectedly (as by a cosmic ray triggering a thyratron switch).
11. Use dark, absorbing, diffuse, fire-resistant targets and backstops when feasible.
12. Design safety into micro-welding and cutting equipment, and similar devices used in miniature work. Because of the increased use of the laser beam in the scribing of integrated circuit chips or trimming of resistors, the use of microscopes or other focusing optics integral to the laser systems is becoming more common.

These high-power laser applications require special attention to their associated hazards. If at all possible, such work should be accomplished in a light-tight or baffled interlocked enclosure to eliminate the requirements for eye protection and other Class IV safety rules. Optics used for viewing the beam and associated areas should either have built-in filters or separate optical paths for intermittent viewing and laser firing.

Control of Associated Hazards

Electrical. The electrical system associated with lasers is the greatest secondary hazard. With some systems, this can be more of a hazard than the laser radiation itself. The power supplies for some lasers use high voltage and should be treated with a great deal of caution. Death from electrocution is a very real possibility in many laser installations. Dangerous or fatal shocks to operators can most easily be avoided by leaving service and repair work to qualified service personnel.

Some high-power pulsed laser systems use high-voltage capacitors to discharge large amounts of energy. Power capacitors are both an electrical and explosion hazard and should be enclosed in special cabinets. Under certain conditions of use, the "discharged" capacitor may retain or restore a significant portion of its original charge.

X-Radiation. Any time that high-voltage equipment is operated at or above 15 KV, the possibility of an X-radiation hazard exists. Any tube, electron gun, or power supply operating at these voltages should be checked with standard radiation protection survey instruments. X-ray exposure levels in the vicinity should be clearly marked with suitable radiation hazards signs indicating the exposure rate to be expected when the laser is operating. Excessive X-radiation levels should be corrected by the installation of additional shielding when necessary. If there is any doubt in your mind as to the existence of an X-radiation hazard associated with your operation, contact the safety officer.

Gas Hazards. Toxic gases and fumes may be encountered in the operation of many lasers. Several different gases are used to operate gas lasers, and each gas has its own toxic characteristics. Some of the lasers that contain toxic gases are mercury-krypton, mercury-zinc, hydrogen cyanide, and fluorine. Toxic gases such as ozone and oxides of nitrogen may be formed when a laser is operated in the ultraviolet region. Toxic fumes can be generated during cutting, drilling, and machining operations or whenever laser target material is vaporized.

High-pressure and liquefied gases are used with many laser systems. Cryogenic gases are used in cooling laser systems and are an auxiliary part of many experimental lasers. Frozen tissue can result from the mishandling of these gases. High-pressure and cryogenic system failures can injure personnel (or damage property) by explosive failure, suffocation, and by contact with intensely cold materials.

Other Hazards. Fire and explosion hazards are possible if the beam is misdirected in the presence of flammable or combustible materials. These materials may be present in the form of gases associated with the equipment, or in the form of material in the path of the beam. Experimental lasers have demonstrated potential fire hazards by igniting paper at distances of several miles. An obvious but important precaution to be taken in the vicinity of lasers is the strict and routine observance of simple "good housekeeping." Flammable materials should be removed from the premises if possible, and if not, they should be stored in cabinets or drawers.

Noise from capacitor bank discharges can be a hearing hazard. Appropriate hearing protection equipment (ear muffs and plugs) can be used to minimize this problem.

IONIZING RADIATION AND RADIOACTIVITY

Ionizing radiation is one of the health hazards least recognizable to the senses. Most hazards possess a physical characteristic such as smell or an appearance that alerts investigators to its presence, but dangerous radiation cannot be identified without the use of instruments or effective labeling.

Ionizing radiation has two main components: short wavelength electromagnetic radiation (X-rays and γ-rays) and subatomic particles (usually, α particles, electrons, positrons, and neutrons).

For the purposes of this chapter, ionization occurs when a neutral atom has an interaction with a passing charged particle, γ photon, proton, or electron that causes one or more of the orbital electrons of the atom to be ejected, leaving the atom charged. The ions formed can be not only the charged atoms but also free electrons, electrically charged parts of atoms, or groups of atoms carrying a charge. Ions react chemically with matter, move in electric fields, recombine, emitting light, and serve as condensation nuclei. These properties can cause tissue damage and serious health problems.

The common sources of ionizing radiation are:

1. Radioactive isotopes that emit α particles, β particles, or γ-ray photons, or some combination
2. X-ray machines
3. High-energy particle accelerators that emit various penetrating charged particles and secondary radiations, as well as produce radioactive materials
4. Reactors that produce neutrons and γ-rays, and also intense radioisotope sources from fission fragments or activation

Quantities and Units

Curie. The basic unit of radioactive decay is the curie, which is defined as 3.7×10^{10} disintegrations per second. This is approximately the rate produced by 1 g of radium in equilibrium with its disintegration products.

Roentgen. The Roentgen (R) is the basic unit of measurement of γ- or X-radiation. It is based on air ionization produced by electron secondaries of X- or γ-radiation.

$$1 \text{ R} = 2.58 \times 10^{-4} \text{ coulomb/kg air}$$

For personal exposure, the R unit is a comparatively large quantity of radiation. It is therefore common to use a fraction of the R as a unit of exposure measurement in radiation protection work. A milliroentgen is one thousandth of a R (abbreviated mR). Many radiation survey and monitoring instruments measure the air ionization produced and are calibrated in mR or mR/hr.

Rad. The rad is the unit of the absorbed dose of ionizing radiation in any material. The absorbed dose is the amount of energy deposited at a point within a material. One rad is 100 ergs absorbed per gram of material at a point of interest in the material. The energy is deposited locally by both ionization and excitation processes occurring at or very near that point. Note that incident photons ionize by means of the secondary electrons they produce. Usually, one specifies the type of materials, since a given radiation type and spectrum will produce different doses in different materials. However, in radiological health work, one normally means rads (tissue) even without specifying tissue.

The importance of the absorbed dose, in rads, is that it is this quantity that produces the effects we see in biological materials and living systems. The most common means of measuring an absorbed dose, however, is in terms of the ionization it produces. The meters used normally read out in R (or mR). Because the rad and R values are numerically close for medium-energy X- or γ-rays, in practice we infer the rad (tissue) dose from the R reading.

The rad can be related to any type of radiation. If the radiation type is medium-energy photons (0.3–3 MeV) and the material is biological tissue or water, then the radiation level that produces an exposure of 1 R will produce 0.97 rad of tissue dose. What this means is that a medium-energy photon beam of X- or γ-rays in which we measure an exposure dose of 1 R will produce 0.97 rad of absorbed dose in soft tissue.

This conversion value varies from 0.94–0.98 over the photon energy range of a few 100 KeV to a few MeV. Because this is so close to one rad (tissue), per R, it is often within the survey monitoring instrument reading error. Hence, in practice the distinction is often lost between tissue-absorbed dose in rad and exposure measured in R. However, note that this approximation is allowable only in the medium-energy range.

Rem. The unit to express the biological dose is the Rem. The rem is not measured directly like the R, but is a computed unit.

The rem is a unit that expresses the biological effect or biological dose in humans for any type of radiation. The origin of the term is "Roentgen-equivalent-man." The biological dose in rems for X-rays or γ-rays is equal to the absorbed dose in rads or the exposure in R (1 γ-ray R $\approx$ 1 rad = 1 rem).

Each of the following is considered to be equivalent to a dose of 1 rem:

A dose of 1 R due to X- or γ-radiation

A dose of 1 rad due to X-, γ-, or β-radiation

A dose of 0.2 rad due to neutrons or high-energy protons

A dose of 0.05 rad due to particles heavier than protons and with sufficient energy to reach the lens of the eye

A dose of approximately 14 million neutrons per cm^2 incident upon the body

LET. As an ionizing radiation particle, an electron or α or even a photon or neutron (which do not ionize directly), passes through material, it deposits some of its energy along its track. It leaves in its wake a trail of ions, which are charged atoms and free electrons, and excited or moving atoms. The amount of energy deposited per unit track length is defined as linear energy transfer, abbreviated LET. The units of LET are energy per unit length, e.g., MeV/cm, or KeV per micron. It takes approximately 33 eV to produce an ion pair in air or light materials such as water. Another commonly used unit for LET is ion pairs per micron in a specified material.

As particle energies vary, LET varies also. Charged particles generally ionize at higher LET, as they slow down at the end of their ranges. Photons of low energies deposit more energy by photo effect, and there is some small variation in equivalent LET as a result.

RBE. In order to be able to compare the effects of different radiation types on the body, relative biological effectiveness, or RBE, is defined as the ratio of the low-LET X- or γ-ray dose to the dose from another radiation that produces the same biological effect, or endpoint. Thus, we can see that RBE depends on both radiation type and the particular biological effect. The endpoints commonly studied involve animal death or loss of specific functions. Note that, by definition, the RBE for low-LET X- or γ-rays is 1.0.

Quality Factor. In order to evaluate hazards of radiation in ordinary situations involving its use, radiobiologists have developed the concept of *quality factor* (QF). QF values have been assigned to all radiation types and LET ranges. The QF value is chosen conservatively to be the highest likely RBE for each LET range for the effects of interest. For example, fast neutron experiments on laboratory animals might indicate an RBE of 6 for some white blood cell effect, an RBE of 4 for hair loss, an RBE of 7 for induction of bone cancer, and other RBE values for other effects. Considering the biological differences between humans and these animals, we might feel that an RBE of 10.5 is likely to be the highest reasonable value for the effects of these fast neutrons on humans. We therefore would assign a QF value of 10.5 to this radiation type (1.0 MeV neutrons). Because of the conservative nature of a QF, it should be used in place of RBE when considering radiation protection.

The biological dose (rem) for any type of radiation may be obtained by multiplying the absorbed dose (rad) by the quality factor.

Biological Effects of Radiation

Radiation damages cells mostly by the activity of free radicals formed in the radiolysis of cell water. Some cells are killed and others experience chromosome and gene changes. The most sensitive part of the cells appears to be the genetic material, or DNA, that controls cell function, enzyme production, and cell reproduction. Generally, cells that are undifferentiated (immature) or are in the process of dividing are more sensitive to radiation. For example, the blood-forming (or hematopoetic) system, consisting of bone marrow and other organs, seems to be particularly sensitive. Also, high LET radiation is more effective in producing cell damage than low LET radiation. A dose of 1 rad (tissue) delivered by 5 MeV α particles will kill more cells or cause more gene mutations than a dose of 1 rad (tissue) delivered by fast neutron-induced protons, which in turn cause more damage than a γ-ray dose of 1 rad (tissue).

Cells that are sublethally injured have some capability for repair. However, some cells are modified by radiation rather than killed outright, and so diseases such as aplastic anemia or leukemia develop, in which blood cells do not perform normal functions or grow in uncontrolled ways.

Acute Radiation Syndrome. When a large number of cells in a body are damaged by a single large exposure in a short time, then the body exhibits acute radiation syndrome. There are three phases to this illness: initially there is nausea and vomiting, then an asymptomatic latent period, followed by the main illness phase. The main illness may involve hematopoetic disturbances, intestinal wall denuding, or central nervous system effects (at very high doses). The symptoms are due to a lack of replacement cells in the blood (white cells, platelets) and intestinal mucosa (villi), and include infections, hemorrhage, diarrhea, and shock. The severity of response depends on the dose.

The dose for which half the human population will die in 30 days (LD 50/30) if untreated is thought to be in the range of 250–400 rem for the whole body. A person receiving over 200 rem requires hospitalization.

Late Effects and Chronic Exposure. Survivors of acute high-dose exposure, and those chronically exposed to low dose rates, may exhibit late effects of radiation. These effects include increased risk of certain types of cancer, notably leukemia, of heritable genetic (sex cell) mutations, cataracts, or a small decrease in life span. These effects are due, in the case of chronic exposures, to accumulated unrepaired cellular injury. Temporary or permanent sterility in males may result from both acute and chronic exposures of the testes. The effects discussed may be produced by either external or internal radiation. The degree of genetic effect caused by radiation exposure is described in Table 11-4. This shows that a radiation worker who receives his 5-rem permissible annual limit each year for 10 years accumulates an increased risk of a dominant genetic mutation roughly equal to or less than his or her chance of being killed in an auto accident during this same period of time. Note that few radiation workers receive the full 5 rem/year. However, recent (1989) NRC studies have shown that the cancer risk from low-level X-ray and γ-radiation is three to four times greater than earlier believed. The risk of developing solid tumor cancers is three times greater and the risk of leukemia is four times greater. There also seems to be a much larger risk of mental retardation among unborn babies than previously thought. All tables in this handbook give information provided before this study.

TABLE 11-4. Magnitude of Genetic Effect

Normal Background
1/17 of all live births have a significant inherited defect
1/100 of all live births have a significant dominant defect
Radiation Effect
1/20,000 to 1/200,000 increased risk from 1 rem
1/400 to 1/4000 increased risk from 50 rem
For Comparison
1/400 average 10-year risk for fatal automobile accidents in the United States

Source: From NIOSH, *Ionizing Radiation 584, Student Manual,* 1981, p. VI-57.

The basis for radiation protection guides, or allowable external exposure levels, involves the late effects of radiation, in particular, carcinogenesis.

Tables 11-5 and 11-6 summarize the types and degree of effects from exposure to ionizing radiation.

SUMMARY OF BIOLOGICAL EFFECTS OF RADIATION

1. There are two types of exposures: single dose (acute) and chronic (long-term, low-rate).

TABLE 11-5. Effect/Mechanisms of Radiation Exposure

Exposure Condition	Somatic Cell	Organism Level	Genetic Cell
Single acute exposure	Stem cell death Mutations	Acute radiation syndrome Sterility Late effects: Cancer Cataracts Aging	Heritable mutations
Chronic exposure	Cell death Mutations Cell repair errors	Sterility Late effects	Heritable mutations

Source: From NIOSH, *Ionizing Radiation 584, Student Manual,* 1981, p. VI-66.

2. There are two basic types of cellular-level responses in somatic (body) cells and genetic (sex) cells.
3. There are three classes of organism-level effects, namely,
   (a) acute radiation syndrome
   (b) temporary or permanent sterility
   (c) late effects (cancer, cataracts, and aging).

Types of Exposure

External Exposure. In external exposures, the source is outside the body as, for example, in a sealed γ-ray source or an X-ray machine. For such exposure, the biological response depends on the total dose and dose rate, as we have seen earlier, and also on the dose distribution in the body. As the body absorbs radiation in its outer layers, there may be some protection afforded to internal organs. Since α particles are of short range, they do not penetrate, and β particles typically irradiate only the skin. Therefore, the external sources of interest are generally X- or γ-rays, and neutrons.

Internal Exposure. When a radioactive material is taken into the body, it can irradiate locally wherever it passes or deposits. This situation leads to what is called internal exposure. The effects of such irradiation are local high doses in some organ(s) or the whole body.

α and β particles can produce large local doses inside the body, and they are taken into account in setting radiation protection guidelines for internal exposure. The concentration of radioactivity in breathable air, drinkable water, or edibles are the quantities that must be controlled, since radioactivity often enters the body via the lungs or mouth.

TABLE 11-6. Summary of Health Risks

	Acute Dose (rem)				Low-Intensity Exposure Dose (rem)			
	100	200	400	800	100	200	400	600
Radiation death	0	0	40%	100%				
Radiation sickness	1%	50%	100%	100%				
Permanent sterility	1%	20%	50%	100%	50%	50%	70%	70%
Temporary sterility	50%	100%	100%	100%	50%	—	—	—
Cataracts	0	0	50%	90%	0	0	0	50%
Leukemia rate increase	6x	12x	24x		3x	5x	10x	15x
Life shortening (avg.)	3–4 yr	4–6 yr	5–11 yr		0–2 yr	3–4 yr	0–8 yr	0–10 yr
Mutation rate increase	2x	3x	4x		2x	3x	4x	5x

Note: These values are estimated based on limited published human effects data.
Source: From NIOSH, *Ionizing Radiation 584, Student Manual,* 1981, p. VI-74.

TABLE 11-7. Maximum Permissible Occupational Dose Equivalent (REM) for People over 18

Organ	Weekly	Quarterly	Annually
Total body, head and trunk, eye lens, gonads, blood-forming organs	0.1	1.25	5
Skin of whole body, thyroid	0.6	7.5	30
Hands, forearms, feet, ankles	1.5	18.75	75

Note: The MPDE for employees under 18 is 10% of the above numbers. The MPDE for expectant mothers and to the fetus shall not exceed 0.5 rem during the entire gestation period.

Source: From FDA, *The Radiation Safety Handbook for Ionizing and Nonionizing Radiation,* Rockville, Md., 1976, p. 13.

Internal exposure is controlled by limiting the surface and airborne contamination of work spaces to the maximum permissible concentration of a particular radionuclide. Allowable airborne concentrations of radioisotopes are tabulated in 20 CFR, Appendix B.

As with external radiation protection guidelines, the basis for setting internal exposure limits involves the late effects of radiation in the critical organs of the body.

Permissible Exposures. Federal regulations require users of ionizing radiation to make every reasonable effort to maintain radiation exposures, and the release of radioactive materials in effluents to unrestricted areas, as low as is reasonably possible.

The maximum permissible dose equivalent (MPDE) to the total body is 100 millirems per week (for adults who are not pregnant). Allowable levels are 10 times less for individuals under 18. Whole-body and extremity-permissible doses are summarized in Table 11-7.

Dose Equivalent Limits for Pupils in Schools. The International Commission on Radiological Protection, in ICRP Publication 26, 1977, concluded that the exposure resulting from demonstrations or experiments with radiation sources is likely to involve a very large number of people and might thus contribute significantly to the general exposure. For this reason, the ICRP reaffirms the detailed recommendations with regard to "school exposure" given in ICRP Publication 13, "Radiation Protection in Schools for Pupils up to the Age of 18 Years." The objective of these recommendations is to achieve a situation whereby the annual dose equivalents received by individual pupils will be most unlikely to exceed one-tenth of the dose equivalent limits recommended for individual members of the public. Table 11-8 lists the recommended annual dose limits.

It is also recommended that no pupil receive more than one-tenth of the above dose equivalent limits in the course of any one demonstration or experiment.

It should be noted that the ICRP intended compliance with these dose equivalent limits to be achieved through the proper, safe design of equipment and experiments rather than by detailed monitoring.

Types of Radiation

X-Rays X-rays are electromagnetic radiation (photons) with wavelengths shorter than ultraviolet light. The wavelength range for X-rays is 0.006–10 nm (0.06–100 Å). The common sources use emission by electron transitions in the inner orbits of heavy atoms that have been bombarded by electrons with energies higher than 15 kV. The shortest wavelengths (highest energy) are called hard X-rays and the longest wavelengths soft X-rays. X-rays were discovered by Roentgen in 1898.

X-rays have the following properties:

1. An ability to penetrate solids of moderate density such as body tissues
2. Absorbed by solids of higher density: bone, lead, barium sulfate
3. Ionize light atoms
4. Expose photographic materials and excite fluorescent materials
5. Can damage and destroy body tissue by both acute and chronic exposure

TABLE 11-8. The Annual Dose Equivalent Limits for Students

Whole body	0.05 rem
Thyroid	0.30 rem
Thyroid of children up to 16 years of age	0.15 rem
Gonads and red bone marrow	0.05 rem
Other organs	0.15 rem

Source: From *Recommended Safety Procedures for the Selection and Use of Demonstration-Type Gas Discharge Devices in Schools,* Canadian Department of National Health and Welfare, Ottawa, 1979, p. 13.

X-ray machines are used for a wide variety of work in the laboratory (e.g., spectrometry, crystallography, cancer therapy, diagnostic medicine, materials testing, food preservation, and teaching) and range from those operating at less than 10 kV to those operating in the megavolt range. Other devices such as high-voltage rectifier tubes also generate X-rays as a byproduct. In general, any vacuum tube with a voltage of more than 10–15 kV should be considered a potential generator of X-rays.

An X-ray-producing device's output increases directly with the atomic number of the target material and with the current through the tube (mA). The output also increases as the square of the operating voltage (kV). At very low operating voltages, the output may increase more rapidly than the square of the voltage because the penetrating power of the radiation also increases, thereby reducing the inherent filtration effect. Normally, the electron beam impinges on a small area of the target so that the X-rays are essentially a point source and follow the inverse-square law. If the beam is of low energy, absorption in the air modifies the inverse-square response, and any radiation scattering causes further modification.

The X-ray tube is normally contained in housing that shields against radiation in all directions, except at an opening left to bring out the useful beam. In most machines, the size of the useful beam can be changed by collimators, and provision is made for inserting filters. It is good practice to use collimators that provide the smallest beam needed for the work.

Common X-Ray Equipment. Field X-ray radiographic inspections are a common use of X-ray equipment. This type of operation requires continuous radiation monitoring to ensure adequate protection. Short-term dosimetry must be worn by X-ray operators. The maximum dose at the perimeter of the hazard area is regulated. Nonoccupationally exposed personnel should not be allowed in the hazard area during exposures. When possible, portable shielding should be used to reduce the size of required perimeters around the hazard area.

The area exposed to the beam is extremely dangerous and should be visually checked for occupancy before each exposure. Scattered radiation should also be considered when hazard area boundaries are determined.

X-ray diffraction and spectroscope units are used in the study of the crystalline structure of matter. They operate at low voltages (20–50 kV) and high amperage power levels to produce massive levels (500,000 R/min) of low-quality radiation. Many old units still in use do not comply with new standards.

When operating these units, remember that most dosimeters have an enhanced response in the photon energy range 20–50 KeV and may be insensitive below about 15 KeV because of shielding in the case.

Demonstration-Type Gas-Discharge Devices in Schools. For many years, gas-discharge tubes have been in widespread use in schools for the experimental treatment of atomic physics. These devices are designed to demonstrate the production, the properties and effects of electrical discharges in gases, and the luminous phenomena that accompanies them. Some discharge tubes are specifically designed to generate X-rays; in others, X-ray emission is incidental to the intended purpose. In its basic form, a gas-discharge tube consists of a partially evacuated, sealed glass tube, containing the gas of interest at a predetermined pressure, and two or more electrodes for applying a high voltage to the enclosed gas.

Gas-discharge tubes are usually classified in accordance with the means used to produce electron emission from the cathode. Discharge tubes in which the electrons are produced by means of thermionic emission are referred to as hot-cathode tubes. By employing a heated filament to produce the electrons, such tubes have been designed to demonstrate the desired effects with operating voltages below 5000 V and without accompanying X-radiation. Several demonstration-type hot-cathode tubes are available commercially. No X-radiation should be emitted from these tubes when operated at voltages within the manufacturer's specified operating range.

The second category of discharge tubes does not employ a heated cathode; the electrons are released from the cold cathode surface by positive ion bombardment. The positive ions are generated by the ionization of gas atoms contained in the tube. Such tubes are classified as cold-cathode tubes. A variety of high-voltage sources are used to operate cold-cathode discharge tubes.

A 1972 survey by the Canadian Radiation Protection Bureau revealed that a number of different types of cold-cathode discharge tubes emitted X-rays incidental to their intended purpose and that the exposure rate increased markedly if the minimum voltage necessary to show the desired effect was exceeded. These tubes will produce X-rays when energized with either forward or reverse potential.

The various types of high-voltage gas-discharge tubes are the cold-cathode X-ray tube, shadow or fluorescence effect tube, heat effect tube, magnetic or deflection effect tube, high- or low-vacuum effect tube. These tubes should not be used for any class laboratory or demonstration work. The risk of X-ray exposure is too high.

α Particles. Alpha particles are helium nuclei emitted spontaneously from some radioactive materials. Their energy lies in the range of 4–8 MeV. They are readily absorbed by most matter, and thus, their energy is dissipated in a very short path, in just a few cm of air or less than 0.005 mm of aluminum.

Because α-radiation from radioactive materials has low penetrating power, it is of concern only when taken into the body (internal exposure). Therefore, α surveys are aimed at detecting contamination that could be transferred to the body by becoming airborne, by ingestion, or, rarely, by absorption into wounds or even through intact skin.

Because of α particles' low penetrating power, instruments must have minimal material between the sensitive volume and the particle source. Thus, the instrument windows are very thin (a few mg/cm^2 at the most), and the probes are designed to permit a close approach to the object surveyed.

The window of an α survey instrument must be held close to the monitored surface. Readings taken 1/4 in. from a surface contaminated with plutonium will be only 70–80% of those at contact with the surface, and readings at 1/2 in. will be only 50–60% of the contact reading. On the other hand, the probe can be contaminated by a surface having loose contamination, and contact must be minimized. The instrument cannot be protected from contamination by a protective cover because this would increase the window thickness and block the α particles.

When surveying, it is advisable to use a check source occasionally to be sure that the instrument is responding properly. The counting efficiency may vary over the surface of the window area. This can affect the reading if the contamination is in a small area. One should, therefore, move the probe about until the maximum reading is obtained. In such a case, one may wish to find the exact point of contamination by using a piece of paper as a mask and moving it to note the effect on instrument response.

β-Rays. β-rays are high-speed electrons that originate in the nucleus. This is in contrast to ordinary electrons, which exist in the orbits around the nucleus. They travel several hundred times the distance of α particles in the air and require, depending on the β particle's energy, a few mm of aluminum to stop them. They are less penetrating than γ-rays.

β emitters pose potential hazards when the body takes in such material or when external radiation penetrates to critical body organs. Protection, therefore, requires measurement and control of both contamination and external radiation levels. β- and γ-radiations are often considered together because most radioactive materials emit both and the techniques for measuring the two are similar. Many times, however, only one emitter is found or the potential exposure from one is much greater than that from the other.

β-γ monitoring instruments can have thicker walls than α instruments, and radiations can be detected farther from the source. β particle contamination is usually surveyed with the instrument window shield open to permit detection of β-radiation. A thin plastic shield can be placed over the instrument to protect it against contamination. It is possible to contaminate a survey monitor, and the past history of the equipment must be considered in assessing control of its future use. Any β or α measurement can be regarded as having the potential to contaminate the monitor. It is a matter of how "fixed" the contamination is.

β-radiation measurements pose another special problem because β penetration is less than that of γ-radiation and is blocked from entering the chamber except through the window. Therefore, those β particles that strike the chamber walls are absorbed, and only those that enter the chamber at the correct angle are detected. Thus, the reading will be low, and the instrument should be regarded as a β-radiation detector (not a survey monitor), requiring special calibrations to measure the dose rate. Note also that the instrument is directional in its β response and measures only radiation coming toward the window.

Another source of error occurs when the meter is held against a surface to obtain a maximum reading called a "contact" reading. In this case, the distribution of the radiation through the sensitive volume will be different from the conditions when the instrument was calibrated. The reading will be lower than the true dose rate in contact with the source. Therefore, the monitor will underestimate dose rates received if the source or article is held in the hands or closely approached. Fixed ratios for this error cannot be given for all situations because the instrument response will depend on the particular source. Ratios can be greater than 1/100 (β measurement vs. true rate). If accurate readings are needed, the instrument should be calibrated for the source of interest.

γ-Rays. γ-rays are a type of electromagnetic radiation (photons). γ-rays have wavelengths shorter than X-rays and thus have higher energy. Whereas X-rays originate in the orbital electron structure of the atom, γ-rays originate in the nucleus.

γ-rays are extremely penetrating. Because of the high energy of the photon, it can cause a lot of damage when absorbed. γ-rays produced by radioactive sources lose energy by interacting with matter in two ways: the photoelectric effect and the Compton effect.

The photoelectric effect is an all-or-none energy loss. The photon imparts all its energy to an orbital electron of some atom.

The Compton effect provides a means of partial energy loss for the incoming photon. Again, the γ-ray appears to interact with an orbital electron of an atom, but only a part of the energy is transferred to the electron, and the γ-ray proceeds with less energy. By this mechanism of interaction, the direction of photons in a beam may be changed, so that scattered radiation may appear around corners and behind "shadow-" type shields.

Background Radiation. Background (natural) radiation in man is created by cosmic radiation, and radioactive substances in the human body and in the environment. Primary cosmic radiation mainly consists of protons and heavier nuclei possessing extraordinarily high energy. Individual particles may have energies up to 10^{18} eV. Interacting with the Earth's atmosphere, these particles penetrate to an altitude of 12 miles above sea level and form secondary hard radiation consisting of mesons, neutrons, and softer components (electrons, γ-rays, etc.).

The intensity of cosmic radiation depends on the geographic location on Earth and increases with height above sea level (Table 11-9). For average latitudes at sea level, the dose on soft tissues on account of cosmic radiation (without a neutron component) is 28 mrem/year, and the neutron component creates an additional dose of 25 mrem/year.

A Radiation Control Program

The purpose of radiation safety is to limit exposure to personnel from ionizing and nonionizing radiation to the lowest amount feasible; however, under no circumstances is exposure to exceed appropriate federal standards.

TABLE 11-9. Power of General Tissue Dose of Cosmic Radiation Depending on Altitude and Latitude, mrem/year

Altitude Above Sea Level, km	Equator	30°	50°
0	35	40	50
1	60	70	90
2	100	130	170
3	170	220	300
4	260	360	500
5	400	530	800
10	1,400	2,300	4,500
15	3,000	5,000	11,000
20	3,500	6,000	14,000

An effective radiation program must provide adequate protection for radiation workers, nonoccupationally exposed personnel, the public, and all associated environs. Depending on the nature and severity of the hazard, physical exams, bioassays, total body counting, personnel monitoring, radiation surveys, environmental monitoring, records, and preplanned emergency procedures may be required to provide this protection.

To implement this concept, a radiation program should ensure:

1. The use of all sources of radiation in a manner that will minimize health and safety risks to the faculty, employees, students, and the public in general
2. The evaluation of radiation hazards to protect personnel
3. A safe work environment in accordance with standards and regulations promulgated by federal and state agencies
4. Immediate investigation of all radiation incidents including overexposures and the establishment of immediate corrective action to prevent their recurrence
5. The control of the release or disposal of all radiation sources or radioactivity
6. The maintenance of an accurate inventory and accountability for all sources of harmful radiation within the organization

The Radiation Safety Officer. A radiation safety officer (RSO) should be designated. He or she is responsible for assuring radiation protection within the confines of that facility. A radiation safety officer's duties include:

1. Maintaining a personnel-monitoring program, including the keeping of internal and external personnel exposure records, notifying individuals and their supervisors of exposures approaching the maximum permissible amounts, and recommending remedial action
2. Instructing personnel in the proper use of personnel-monitoring devices
3. Performing or supervising periodic or special radiation surveys and monitoring of all radionuclide and radiation facilities, and maintaining records of all such surveys
4. Ordering, receiving, storing, processing, and dispensing all radionuclides and maintaining pertinent records

5. Inspecting facilities and equipment to be used in conjunction with radionuclides and/or radiation-producing equipment to ensure appropriate radiation safety features
6. Furnishing consulting services to personnel at all levels of responsibility on all aspects of radiation protection
7. Supervising and coordinating the waste disposal program, including the keeping of waste storage and disposal records
8. Storing all radioactive materials not in use
9. Performing leak tests on all sealed sources and keeping records of such tests
10. Obtaining and maintaining current copies of all federal and state regulations pertaining to radiation safety, and having immediately available copies of all NRC (AEC) licenses

Individual Responsibility for Radiation Protection. Each individual who is designated as a user or operator of a radiation-producing machine or who has contact with any radioactive material is responsible for:

1. Keeping his or her exposure to radiation as low as possible, and specifically below the maximum permissible dose equivalent as listed in Tables 11-8 and 11-9.
2. Wearing the prescribed personnel-monitoring equipment such as TLD's (thermoluminescent dosimeters), film badges, and pocket dosimeters in radiation areas.
3. Each individual user should utilize all appropriate protective measures including the following:
   (a) Wear protective clothing whenever contamination is possible.
   (b) Wear gloves and, when necessary, adequate respiratory protection devices.
   (c) Use pipet filling devices. Never pipet radioactive liquids by mouth.
   (d) Work with radionuclides within the confines of an exhaust hood or glove box, unless approval has been granted by the radiation safety committee (RSC) for working in the open.
4. Each user should survey his or her hands, shoes, and body for contamination before leaving radiation areas.
5. Avoid smoking, drinking, or eating in areas where radioactive materials are present. Smoking or eating may be permitted in an office adjacent to such an area when it has been demonstrated that the office is free of contamination.
6. Maintain good personal hygiene.
   (a) Keep fingernails short and clean.
   (b) Do not work with radioactive materials without protective gloves if there is a break in the skin below the wrist.
   (c) Wash hands and arms thoroughly before handling any object that will contact the mouth, nose, or eyes.
7. Survey the immediate areas of hoods, benches, and so forth, during and after the use of radioactive materials. Any contamination should be removed immediately. If such removal is not possible, the area should be clearly marked and the RSO notified.
8. Keep the area containing radioactive materials neat and clean. The work area should be free of equipment and materials not required for the immediate procedure.
9. Store or transport materials in appropriate containers, preferably double containers, to prevent breakage or spillage and to insure adequate shielding.
10. Keep work surfaces covered with absorbent material and employ stainless steel trays or pans to limit and collect spillage in case of an accident.
11. Label and isolate radioactive waste and equipment, such as glassware, used for radioactive materials. Once equipment is used for radioactive substances, it should not be used for other work or sent from the area to cleaning facilities, repair shops, or to surplus, until it is demonstrated to be free of contamination.
12. Report all accidental releases, inhalation, ingestion, or injury involving radioactive materials to the supervisor and RSO and carry out recommended measures.
13. Undertake decontamination procedures when necessary and take the appropriate steps to prevent any additional spread of contamination.
14. Do not use refrigerators jointly for foods or film and radioactive materials.

Radiation Areas. A "radiation area" is defined by OSHA as any area accessible to personnel in which radiation exists at such levels that a major portion of the body or critical organ could receive in any 1 h a dose in excess of 5 mrem, or in any 5 consecutive days a dose in excess of 100 mrem. Each radiation area must be conspicuously posted with a sign or signs bearing the radiation caution symbol and the words:

CAUTION (or **DANGER**)
RADIATION AREA

A "high-radiation area" means any area accessible to personnel in which radiation exists at such levels that a major portion of the body or critical organ could receive in any 1 h a dose in excess of 100 mrem.

1. Each high-radiation area must be conspicuously posted with a sign or signs bearing the radiation symbol and the words:

   CAUTION (or **DANGER**)
   RADIATION AREA

2. Each high-radiation area shall be equipped with a control device that shall either cause the level of radiation to be reduced so that an individual would not receive a dose of 100 mrem in 1 h upon entry into the area, or shall energize a conspicuously visible or audible alarm in a manner that the individual entering the area shall be aware of the fact that the potential exists for a dose of 100 mrems or greater.

An "airborne radioactivity area" is any area where there is airborne radioactive material that exceeds levels specified in 10 CFR, Part 20.203 (D) (1) (ii). Each airborne radioactivity area must be conspicuously posted with a sign or signs bearing the radiation caution symbol and the words:

CAUTION (or **DANGER**)
AIRBORNE RADIOACTIVITY AREA

Unrestricted areas also have radiation limits, if an individual is continuously present in an area where he or she could receive a dose in excess of 2 mrems in any 1 h or a dose in excess of 100 mrems in any 7 consecutive days.

As with other forms of radioactivity, the Nuclear Regulatory Commission requires that exposures for people under 18 years be much less than those for adults (usually, 10%). Thus, for schools the definitions of radiation areas and unrestricted areas should be more strict.

Each area or room in which licensed material is used or stored and which contains any radioactive material (other than natural uranium or thorium) in an amount exceeding 10 times the quantity of such materials as specified in 10 CFR, Part 20, Appendix C must be conspicuously posted with a sign or signs bearing the radiation caution symbol and the words:

CAUTION
RADIOACTIVE MATERIALS

External Radiation Control Methods. External radiation exposure is usually from the following sources: X-rays, γ-rays, β-rays, or neutrons. The exposure is normally limited by controlling exposure time, the distance from the source, and placing shielding to absorb the radiation.

Personnel monitoring is another one of the methods used to control external radiation. It involves estimation of an individual's exposure through the use of instruments carried on that person or through the analysis of body fluids or tissue. Such monitoring is historical in that it indicates exposures that have already occurred. The results are useful in indicating work where additional attention must be given to control measures and in providing some measure of individual radiation exposure. The choice of personnel-monitoring devices must be compatible with the types and energy of radiations being measured. For example, external monitoring devices would not be at all applicable for a low-energy carbon-14 exposure evaluation. The proper use of film badges, pocket dosimeters, and personnel alarm devices will provide a wide range of options for external radiation exposure evaluation.

These controls will result in lower overall exposure values for all concerned individuals. This, of course, is the ultimate goal of a radiation protection program: to hold personnel exposures to as low a value above the background exposure as practical.

Operating Radiation-Producing Machines. Radiation-producing machines must be operated only by designated trained personnel. They must conform to the following rules:

1. Areas in which radiation-producing machines are located or are being used shall be posted with the characteristic "Caution Radiation" or "Caution X-rays" sign to warn unauthorized personnel from entering the radiation area. In addition, the controls shall have the following decal: "Caution Radiation—This equipment produces radiation when energized." In certain instances, other precautions, such as locking the entrances to the room or the use of automatic safety devices, may be required.
2. The radiation-producing machine shall be disconnected from its power source or locked when not in actual use.
3. The operator must clear the primary beam and scatter exposure areas of all personnel before

operating the machine and assure that all essential personnel are adequately shielded.

4. Personnel-monitoring devices must be worn by the operator and all others present during the exposure.
5. The operator must never expose him- or herself to the direct beam of the machine and must not enter the exposure (or irradiation) room while the machine is in operation unless adequately shielded.
6. Whenever technically feasible, the primary beam will be directed downward to a target. Under no circumstances will it ever be directed toward an interior wall or ceiling in the absence of primary beam shielding.
7. All personnel shall observe all restrictions on the use of the machine recommended by the responsible people (the RSO and/or RSC).
8. The structural shielding requirements for any new installation or for any existing one in which changes are contemplated should be discussed with the RSO.

Radioisotopes

Because of their severe potential health hazard, radioactive materials are controlled strictly by the Federal Nuclear Regulatory Commission and by state agencies. The regulations promulgated by these authorities are the most useful method of practically determining the potential health hazard of a given radioactive source.

The regulatory agencies issue radioactive material licenses to organizations wishing to use radioisotopes. The license specifies the maximum radioactivity that may be possessed at one time and the largest single source permissible for each isotope. The individual responsible for control of the radioactive material within the licensed organization must be designated and qualified by training and experience to properly handle radioactive materials.

Materials with very low (but still useful) levels of radioactivity may be used by unlicensed organizations as "exempt" materials. Radioactive sources determined as exempt by the regulations are generally regarded as posing no health hazard to the user and require no license for possession.

Table 11-10 details typical exempt quantities for a variety of isotopes. The permissible quantity for each isotope varies according to its radiotoxicity. You should check to determine whether these figures are applicable to your particular locality, as the regulations vary from state to state. However, as an indicator of relative health hazards, these figures are an effective guideline.

Since radiation health hazards depend on the quantity of radiation received by the human body, the regulations recognize that multiples of exempt sources that exceed the limits in Table 11-10 do pose a potential health hazard and require licensing by the appropriate authority.

Within the school science laboratory, exempt quantities are nearly always used. The head of the department should determine if sufficient exempt quantities are possessed for a license to be required.

Even if no license is required, the students will benefit from a program that uses proper controls for the handling of radioactive materials. The training will prepare them for the many industrial and medical circumstances where strict controls over the storage, issuance, and use of isotopes are required.

A pattern of operation for the control of radioisotopes is described below. A similar procedure is recommended whenever radioisotope sources are possessed. This pattern may also be used as a model for developing procedures to comply with state or federal licensing requirements.

General Principles of Radiation Protection. When radioactive materials are used, there must be suitable storage and protective equipment available. Because of the hazard presented by radioactive material, it should be stored in a locked cabinet or compart-

TABLE 11-10. Some Radioisotopes Available in License-Exempt Quantities

Material	Type of Radiation	μC	Half-Life
Carbon-14	β	100	5,570 years
Cesium-137	β and γ	10	30 years
Cobalt-60	β and γ	1	5.3 years
Polonium-210	α	0.1	138 days
Strontium-90	β	0.1	29 years

Source: From Gerlovich, J. A. and Downs, G. E., *Better Science Through Safety,* Iowa State University Press, Ames, Iowa, 1981, p. 72.

ment. If the activity of the radioactive substance is high enough, a special lead-lined, key-operated box may be necessary for adequate protection.

For routine laboratory work with radioactive materials, the laboratory should maintain a supply of shielding materials including lead sheet, lead bricks, and cement blocks. Personal protective equipment—lead-filled gloves and aprons—may also be necessary if maximum protection is desired.

A radiation-monitoring instrument capable of measuring the type of radiation emitted by the isotope is also essential, and a trained radiation safety officer is required to maintain safe conditions.

Setting up a Program for the Control of Radioisotopes.

1. A responsible member of the staff should be appointed as the RSO for the school. He or she will be responsible for all aspects of procuring, storing, handling, and disposing of radioactive materials and for maintaining the exempt-quantity status; this individual also will ensure compliance with the terms of licensing.
2. A leaded safe should be provided for the permanent storage of all sources. The safe should be identified with a radiation label and should be locked at all times.
3. A log book should be maintained for signing isotopes in and out of the safe.
4. A source record should be maintained on each isotope source possessed by the department.
5. A copy of the departmental radioisotope regulations should be posted at the safe, along with the log and a copy of the license.
6. Isotopes should be transported from the safe in lead blocks. The isotopes should be returned at all intermediate stages of the experiment when they are not in use. The lead block and isotopes should carry the radiation sticker.
7. All sources with a half-life greater than 30 days should be leak-tested prior to initial use and every 6 months thereafter. Sources with a leakage greater than 0.005 μC should be rejected.
8. Keep a record of radioisotope purchases, date of receipt, quantities used, user names, and spills.

Rules for the Control of Radioisotopes.

PROCUREMENT OF RADIOACTIVE MATERIALS

1. Before staff member attempts to procure any radioactive material, he or she must notify the RSO of that intention. The RSO will determine the need for the material, as well as any unusual radiation hazards it might present.
2. All purchase orders for radioactive materials must be initialed by the RSO or deputy RSO before leaving the premises.

RADIOISOTOPE RECEIVING PROCEDURES

1. Upon receipt of any radioactive materials, a radioisotope source record form for such material should be filled out and the RSO or RSO deputy should be notified by the person who has received that material.
2. The RSO should wipe test the package for contamination. Packages should be opened carefully by the RSO.
3. As the package is opened, the condition of the packing material should be observed and any discolorations noted. The packing material should be checked for radioactive contamination.
4. If any contamination, leakage, or shortages are observed, notify the vendor immediately.

HANDLING OF RADIOACTIVE MATERIALS

1. Before removing a radioactive material from the safe, notify the RSO.
2. All radioactive material users must fill out the removal form prior to removing material from the radioactive material storage safe.
3. All radioactive material removed from the storage safe must be returned to the safe at the end of the experiment or end of the day.
4. All radioactive sources removed from the safe must be transported and stored in lead storage containers designed for that purpose.
5. Use a radiation monitor that can detect the type of radiation emitted by the isotope. A good γ-ray detector may be very poor at detecting β-rays.
6. Work over a tray lined with absorbent paper when using radioactive liquids. The bench-top should be covered with absorbent paper with a nonpermeable liner underneath.

7. Wipe up spills immediately.
8. Wear disposable gloves when handling liquid radioactive sources. Check the gloves for contamination frequently.
9. Use forceps or tongs to handle radioactive materials.
10. Never pipet radioactive liquids by mouth.
11. Use a labeled container for radioactive wastes, one set aside for that purpose.
12. Radioactive materials should never be stored near reactive or corrosive chemicals. Keep all acids away from sodium iodide containing iodine-131 and carbonates containing carbon-14 to avoid the generation of radioactive gases. Work in a fume hood if risk of radioactive gas generation exists.
13. Before leaving the work area, check hands, shoes, and clothing for contamination.
14. NRC-approved labels should be used on all glassware containing radioactive materials. Labels should indicate the radioisotope, the activity level, and the date.
15. Keep equipment, glassware, and materials that are not involved in the immediate operation away from the area.
16. Use sealed radioactive sources instead of liquid solutions whenever possible.

Containers. Any container that is used to transport and store or is contaminated with a quantity of licensed material (other than natural uranium or thorium) greater than the quantities specified in 10 CFR, Part 20 must have a clearly visible label bearing the radiation caution symbol and the words, "Caution, Radioactive Material."

Laboratory containers, such as beakers, flasks, and test tubes, used transiently in laboratory procedures do not require labels when the user is present. When such containers are to be left unattended for periods of 8 h or more and contain materials in quantities greater than those specified in 10 CFR, Part 20, they must be labeled. When these containers are used for storage, the labels required by this section must state also the quantities and kinds of radioactive materials in the containers and the date the quantities were measured.

Gas Chromatography Detectors Using Radioisotopes. All gas chromatographs that use radioactive materials should be regulated as follows:

1. Purchased radioactive foils to be used in gas chromatography cells should be shipped to the RSO or the individual designated responsible.
2. Each cell containing a foil should be registered by the RSO or the individual designated responsible and assigned a number. Maintain a file describing the type of source and its location.
3. Each cell containing a radioactive foil should have a label bearing the radiation caution symbol and the words, "Caution Radioactive materials" and the identity and activity of the radioactive material.
4. The radioactive foil should not be removed from its identifying cell, except for cleaning, and should not be transferred to other cells.
5. The following information should be visible in a conspicuous location on the exterior of the gas chromatograph:

   This equipment contains a radioactive source registered with the RSO as required by license from the NRC. Notify the RSO before removing the source from this location or upon any change in custodial responsibility.

6. Gas chromatographs utilizing radioactive sources must be vented through plastic tubing into a chemical hood or room exhaust to avoid the contamination of work areas from the release of radioactive tagged samples introduced into the system or from the accidental overheating of radioactive foils in the cells.
7. Radioactive cells and foils should have periodic leak tests and records should be kept on these tests.
8. Work on radioactive cells such as cleaning should be performed in a hood with absorbent material covering the work space. Gloves should be worn during cleaning.
9. Liquids generated during cleaning may be disposed of into a sanitary sewer with large quantities of water.
10. Do not exceed the temperature limits specified by the manufacturer.

Waste Disposal. Radioactive wastes must be disposed of in accordance with regulations. This means particularly that solid wastes may not be placed in the standard waste containers to be collected by maintenance personnel and that liquid wastes should not be discharged into the sewer without permission from the RSO.

The NRC specifies maximum releases of specific radioactive materials in 10 CFR, Part 20.106 and Parts 20.301–20.306. When there is a combination of isotopes in known amounts, the limit for the combination should be derived as follows: Determine, for each isotope in the combination, the ratio between the quantity present in the combination and the limit otherwise established for the specific isotope when not in combination. The sum of such ratios for all the isotopes in the combination may not exceed 1.

The NRC requires that the gross quantity of licensed and other radioactive material, excluding hydrogen-3 and carbon-14, released into the sewage system by a license does not exceed 1 curie per year. The quantities of hydrogen-3 and carbon-14 released into the sanitary sewage system may not exceed 5 curies per year for hydrogen-3 and 1 curie per year for carbon-14.

Any licensee may dispose of the following licensed material without regard to its radioactivity:

1. 0.05 μC or less of hydrogen-3 or carbon-14, per gram of medium, used for liquid scintillation counting.
2. 0.05 μC or less of hydrogen-3 or carbon-14, per gram of animal tissue averaged over the weight of the entire animal.

If the radionuclide has a short half-life, it can be stored until the activity has decreased to the point where it can be safely handled by normal waste disposal techniques. A short half-life is anything less than 40 days.

Waste Containers. To insure that solid and liquid wastes are kept separate, each laboratory with radioactive waste should be equipped with at least one container for solid dry waste and one for liquid waste.

Radioactive Liquid Disposal. Liquid radioactive materials may be disposed of by pouring them down the drain only if the applicable state and federal regulations are adhered to. Before disposing of such substances in a sewer, get permission from the RSO. Whenever possible, liquid wastes should be poured into a heavy-duty plastic bag partially filled with an absorbent material, the bag then sealed and placed in a sealable metal can. Store the can in the metal drum designated for that purpose. Whenever possible, liquid wastes should be solidified and disposed of as solid waste.

Solid Wastes. Dispose of solid radioactive materials in the manner described for liquid sources above. Measure the level of radiation emitted by each source at a disposal time and record this information on the source form. Radioactive solid waste should be disposed of only by contract with a licensed commercial firm. Packaging, labeling, and quantities of activity permitted per package should be in accordance with the DOT hazardous materials regulations.

Liquid Scintillation Vials. All liquid scintillation vials (except as specified above) must be disposed of as radioactive waste. Check with the disposal firm for proper packaging techniques.

Animal Carcasses. Animal carcasses containing radionuclides should not be disposed of in a general-purpose incinerator. Small animals containing radionuclides should be placed in sealed plastic bags, tagged with the date, the name of the isotope used, and the amount in millicuries. The carcass will then be stored in an appropriately labeled freezer until the isotope decays to background levels or arrangements have been made for the disposal of it as radioactive waste with a commercial firm.

All stages of the disposal procedure must be monitored by the RSO. If, at any time, detectable radiation at the outside surface of the storage drum reaches a level such that any person working near the surface of the drum for 8 h a day, 5 days a week would receive a radiation dose in excess of 0.30 rems per calendar quarter, the drum should be disposed of as soon as possible.

Contamination Guidelines. Suggested surface contamination guidelines for restricted and unrestricted areas within a laboratory are listed in Table 11-11.

Decontamination and Emergency Procedures. Emergencies generally will be in the nature of spills, fires, or explosions, by which radioactive materials can be dispersed or released. Successful decontamination calls for planned action. A spur-of-the-moment action or attempt at decontamination can cause more harm than good. The person responsible for the spill in a contamination accident will usually take the first steps in bringing the situation under control. Those persons responsible for a spill should, unless physically unable, be responsible for all decontamination of the area under the direction or supervision of the RSO. The first consideration should be personnel safety; persons not involved in the accident must leave the area.

TABLE 11-11. Surface Contamination Guidelines

Type of Radiation	Total dpm*/100 cm²	(Removable Smear) Surface Contamination	Total mR/h at 1 in.
α	25,000 max.	500	—
	5,000 avg.	—	—
β-γ	50,000 max.	5,000	1.0
	10,000 avg.	—	—
		Unrestricted Areas	
α	1,000 max.	100	—
β-γ	10,000 max	1,000	0.5
		Skin and Personal Clothing	
α	500 max.	none detectable	—
β-γ	1,000 max	none detectable	0.1
		Release of Material	
α	2,500 max.	none detectable	—
	500 avg.	none detectable	0.1
β-γ	10,000 max.	100	0.5
	2,000 avg.	200	0.2

*dpm = disintegrations per minute.

Note: These guidelines are subject to the following conditions and interpretations:

1. These limits are to be used as guides, and in practice, professional judgment should be used by the RSO to determine the acceptability of the actual contamination.
2. Although it is felt that the recommended values should not result in a health hazard, good radiation protection practice dictates that a reasonable effort be made to keep contamination levels below these values.
3. Compliance with contamination guides should not be used as evidence that the internal exposure of persons to sources of radiation is within the prescribed standards. Biological sampling or whole-body counting should be used to ascertain suspected internal exposure.

Source: From FDA, *The Radiation Safety Handbook for Ionizing and Nonionizing Radiation,* Rockville, Md., 1976, p. 14.

In case of emergency, the following general procedures should be followed:

1. In the event of a fire, explosion, spill, or hazardous malfunction, notify all persons to evacuate the area at once.
2. Prevent the spread of contamination by shutting off ventilation fans, heating, and air-conditioning equipment if airborne contamination is possible. Apply absorbent material in the case of liquids and rope off or barricade the area.
3. Immediately notify the RSO and your immediate supervisor.
4. Monitor all persons involved in the emergency or control action.
5. Make full use of monitoring instruments and available assistance. Each step of the decontamination should be monitored. One person should remain uncontaminated to operate instruments and perform other monitoring. When the instruments become contaminated, further progress is impaired. Protective clothing, footwear, gloves, and respiratory equipment should be used as needed.
6. Following the emergency, monitor the area and determine the protective devices necessary for safe decontamination. The RSO will be available for this determination.
7. The responsible supervisor should prepare a complete history of the emergency and subsequent activity, including corrective and preventive actions taken.

The following lists are more specific emergency procedures for the most common laboratory accidents.

Minor Spills.

1. Remove all unnecessary personnel from the immediate vicinity of the spill.

2. Attempt to contain the spill and absorb the liquid with large sheets of filter paper stored for that purpose.
3. Turn off all air-moving equipment (e.g., fans and air conditioners).
4. Make sure that no one leaves the area with contaminated footwear or clothing.
5. Don the appropriate protective clothing designated for radiation control: rubber gloves, plastic boots, lab coat.
6. Decontaminate.
7. Monitor all persons involved in the spill and cleanup.
8. Permit no one to resume work in the area until a survey is made, and the approval of the RSO is obtained.
9. Prepare a history of the accident and subsequent cleanup for permanent inclusion in the department records.

Accidents Involving Personnel Injury and Radioactive Materials.

1. Wash minor wounds immediately under running water, while spreading the edges of the wound.
2. Report all radiation accidents involving wounds, overexposure, ingestion, and inhalation to the RSO as soon as possible.
3. Call a physician at once who is qualified to treat radiation injuries.
4. Permit no person involved in a radiation injury to return to school without the approval of the RSO and the attending physician.
5. Prepare a complete history of the accident and subsequent related activity for the RSO's records.

Fires or Other Major Emergencies.

1. Notify all other persons in the room and building at once.
2. Attempt to extinguish fires if a radiation hazard is not immediately present.
3. Notify the RSO.
4. Notify the fire department and other local plant safety personnel.
5. Govern firefighting or other emergency activities by the restrictions of the RSO.
6. Following the emergency, monitor the area and determine the protective devices necessary for safe decontamination.
7. Decontaminate under the supervision of the RSO.
8. Permit no person to resume work without the approval of the RSO.
9. Monitor all persons involved in combating the emergency.
10. Prepare a complete history of the emergency and subsequent activities for the RSO's records.

General Procedures for Personnel Decontamination. Ordinarily, the same procedures used for personal cleanliness will suffice to remove radioactive contaminants from the skin, but the specific method selected will depend on the form (grease, oil, and so forth) of the deposited contamination. Soap, sequestering agents, and detergents normally remove more than 99% of contaminants. If it is necessary to remove the remainder, chemicals can be used on the outer layers of contaminated skin. Because of the risk of injury to skin surfaces, these chemicals (citric acid, potassium permanganate, sodium bisulfite, and so forth) should be applied with caution, preferably under medical supervision. Lanolin-based creams are used to offset local irritations of skin surfaces after decontamination.

Remove any clothing or equipment found to be contaminated before determining the levels of skin contamination.

Decontaminate any areas of the body found to have significantly higher contamination than surrounding areas. This spot cleaning is necessary to prevent the spread of contamination to clean areas of the body that might occur in showering.

If contamination is all over the body surfaces, a very thorough shower is necessary. Special attention should be paid to such areas as the hair, the hands, and the fingernails. After showering and monitoring, the residual contamination can be removed by spot cleaning.

Avoid the prolonged use of any one method of decontamination. Repeated ineffective decontamination methods may irritate the skin and thus hamper the success of more suitable decontamination procedures. No one chemical treatment is known to be appropriate for all of the elements with which one may become contaminated.

Avoid the use of organic solvents. Organic solvents may increase the probability of radioactive materials penetrating through the pores of the skin. Oxalic acid is a poisonous compound and is not to be used under any circumstances.

A Specific Procedure for Hand Decontamination. Wash the skin thoroughly with lava soap and water for 2–3 min. Special attention should be paid to areas between the fingers and around the fingernails. A soft brush may also be used, but be careful not to erode or break the skin. Repeat the procedure if monitoring indicates contamination remaining on the skin in amounts above tolerance and if a significant amount of contamination was removed.

When the normal handwash is not effective, apply a sequestrant-detergent liquid mixture (a 5% water solution with a mixture of 30% Tide, 65% Calgon, and 5% Carbose). Rinse with water. Repeat the procedure if the results prove encouraging.

If the liquid detergent mixture is not effective, apply a sequestrant-detergent cream (a 4% Carbose, 3% Versene, 8% Tide, and 85% water mixture). Rub thoroughly into the skin for approximately 1 min. Rinse with water. Repeat the treatment as long as the results show that the contaminant is being removed.

Another method is to place the contaminated hand or hands in surgical gloves with cuffs taped tightly to the wrist. The resultant perspiration after 1 or 2 h of wear will usually wash the contamination from the pores of the skin. *Caution:* When removing the gloves, be sure that they are turned inside out.

Other Decontamination Procedures.

CLOTHING

1. If levels permit, wash clothing in a normal fashion. The recommended maximum permissible level for this action is 0.1 mrem/h.
2. If clothing is too heavily contaminated, place in a heavy-duty plastic bag and then in an isolated steel drum as a prelude to disposal.

FLOORS, WALLS, BENCHES

1. Decontaminate first by scrubbing the floor with a detergent solution.
2. If (1) does not work, scrub with a 25% solution of Contrad 70. Rinse with 1% hydrochloric acid.
3. If further decontamination is necessary, wash the floor with 20% hydrochloric acid.

METAL AND PLASTIC TOOLS

1. Wash with a dilute (1%) solution of nitric acid or a 10% solution of sodium citrate.
2. Or wash with a 10% solution of Contrad 70, followed by a 1% hydrochloric acid rinse.

POROUS SURFACES

Use radioactive materials on nonporous surfaces only; porous surfaces are extremely difficult to decontaminate.

GLASSWARE

1. Soak glassware in a 25% solution of Contrad 70 or its equivalent for 24 h and rinse with water, followed by a 1% hydrochloric acid rinse.
2. Soak glassware in concentrated nitric acid for 24 h and rinse.
3. If (1) and (2) do not work, soak glassware in 50% hydrochloric acid and rinse.
4. If (1–3) fail, dispose of glassware in the proper manner.

SUGGESTED READINGS

American National Standards Institute, Z136.1, American National Standard for the Safe Use of Lasers, New York. Use the latest edition.

Chang, I., *Professional Safety,* Nov. American Society of Safety Engineers, Desplaines, Ill., 1986, pp. 50–53.

NIOSH, *Ionizing Radiation 584—Student Manual.* Available from the National Technical Information Service, Washington, D.C., 1981.

NIOSH, *Nonionizing Radiation 583.* Available from the National Technical Information Service, Washington, D.C., 1977.

International Commission on Radiological Protection, "Recommendations of the International Commission on Radiation Protection," ICRP Publication 26, *Annuals of the ICRP,* Vol. 1, No. 3, Didcot Oxfordshire, England, 1977.

International Commission on Radiological Protection, "Radiation Protection in Schools for Pupils up to the Age of 18 Years," ICRP Publication 13, Didcot Oxfordshire, England, 1968.

Title 29, Code of Federal Regulations, Part 1910.96. Ionizing Radiation.

Title 29, Code of Federal Regulations, Part 1910.97. Non-ionizing Radiation.

Title 10, Code of Federal Regulations, Part 20. Standards for Protection Against Radiation.

CFR titles (chapters) are available from the U.S. Government Printing Office, Washington, D.C.; each title is revised annually.

12
SPILLS AND FIRES

CHEMICAL SPILLS

Experience has shown that the accidental release of hazardous substances is a common enough occurrence to require preplanning for procedures that will minimize the exposure of personnel and property. Such procedures may range from keeping a sponge mop and bucket available to having an emergency spill-response team, complete with protective apparel, safety equipment, and materials to contain, confine, dissipate, and clean up the spill.

In any event, there should be supplies and equipment on hand to deal with the spill, consistent with the hazards and quantities of the spilled substance. These cleanup supplies should include neutralizing agents (such as sodium carbonate and sodium bisulfate) and absorbents (such as vermiculite and sand). Floor-drying compounds such as those used by garages are cheap and effective spill absorbents. Paper towels and sponges may also be used as absorbent-type cleanup aids, although this should be done cautiously. For example, paper towels used to clean up a spilled oxidizer may later ignite, and appropriate gloves should be worn when wiping highly toxic materials with paper towels. Also, when a spilled flammable solvent is absorbed in vermiculite or sand, the surface area of the solvent is increased and the resultant solid is highly flammable and gives off flammable vapors, and thus, it must be properly contained or removed to a safe place. The same applies to the use of paper towels. Do not leave paper towels or other materials used to clean up a spill in open trash cans in the work area. Dispose of them properly.

Commercial spill kits are available that have instructions, absorbents (one of these, SOLUSORB[RT], not only absorbs spilled liquid but also reduces its vapor pressure to a relatively safe level and thus reduces the fire hazard), reactants, and protective equipment. These kits may be located strategically around work areas much as fire extinguishers are. Home-made spill kits can be easily constructed by students and can be an interesting student project.

Preplanning for spills should include consideration of the following factors:

1. Potential location of the release (e.g., outdoors versus indoors; in a laboratory, corridor, or storage area, on a table, in a hood, or on the floor).
2. The quantities of material that might be released and whether the substance is a piped material or a compressed gas.
3. Chemical and physical properties of the material (e.g., its physical state, vapor pressure, and air or water reactivity).
4. Hazardous properties of the material (its toxicity, corrosiveness, and flammability).
5. The types of personal protective equipment that might be needed.

If a spill does occur, the following general procedures may be used, but should be tailored to individual needs. Information more specific to classes of chemicals is contained in the following sections.

1. Attend to any persons who may have been contaminated. Remove contaminated clothing. Flush skin with water. Afterwards wash skin with soap and water.
2. Notify persons in the immediate area about the spill.
3. Evacuate all nonessential personnel from the spill area.
4. If the spilled material is flammable, extinguish all flames and turn off all spark-producing equipment (e.g., brush-type dc motors).

5. Avoid breathing vapors of the spilled material; if necessary, use a respirator.
6. Leave on or establish exhaust ventilation if it is safe to do so.
7. Clean up spill with the most appropriate technique.
8. Wear appropriate protective gear during a cleanup.
9. Notify the safety coordinator if a regulated substance is involved.

The following materials are very hazardous by reason of toxicity or reactivity and should be cleaned up by experienced personnel only:

Aromatic amines
Bromine
Carbon disulfide
Cyanides
Ethers
Nitro compounds
Organic halides

The Chemical Entries section in *Academic Laboratory Chemical Hazards Guidebook* contains specific spill cleanup information for several hundred chemicals.

Spilled Solids

Most solid spills can be brushed into a dustpan and disposed of in appropriate solid-waste containers, but care must be exercised to avoid reactive combinations. Generally, sweep spilled solids of low toxicity into a dustpan and place them in a solid-waste container for disposal. Additional precautions such as the use of a vacuum cleaner equipped with a HEPA filter may be necessary when cleaning up spills of more highly toxic solids.

Nonflammable Liquids

1. Confine or contain the spill to a small area. Do not let it spread.
2. For small quantities of inorganic acids or bases (<100 mL), use a neutralizing agent or an absorbent mixture. For acids use sodium bicarbonate powder or trisodium phosphate plus sand. For alkalies use boric acid powder. A dustpan and brush should be used to clean up the adsorbent. Rubber gloves should be worn. For larger amounts of inorganic acids and bases, flush with large amounts of water (provided that the water will not cause additional damage). Flooding is not recommended in storerooms where violent spattering may cause additional hazards or in areas where water-reactive chemicals may be present.
3. For oxidizers use sodium thiosulfate plus soda ash or diatomaceous earth as an absorbent.
4. For small quantities of other materials, absorb the spill with a nonreactive material [such as vermiculite, Oil-Dri, Zorb-All, towels, or dry sand (although it is less effective)]. *Caution:* Vermiculite and some other absorbents create a slipping hazard when wet. Sodium bicarbonate is an effective spill control agent for neutralizing and immobilizing small spills.
5. Nonflammable organic liquids that are also nontoxic may be treated in the same way as flammable spills. However, if the spill is 100 mL or less, paper towel absorbents may be used. Sand- and clay-based adsorbents may also be used as absorbents.
6. Mop up any remaining spill, wringing out the mop in a pail equipped with rollers.
7. Carefully pick up and clean any cartons or bottles that have been splashed or immersed.
8. Vacuum the area with a vacuum cleaner approved for the material involved, remembering that the exhaust of a vacuum cleaner can create aerosols and thus should be vented to a hood or through a filter.
9. If the nonflammable spilled material is extremely volatile, let it evaporate and be exhausted by the mechanical ventilation system.
10. Dispose of residues according to safe disposal procedures.

Flammable Solvents

Spilling a flammable organic solvent presents a dual hazard. The material may only be flammable, but often it is both flammable and toxic. First consider the flammable aspects. Remember to first turn off all sources of ignition. Breathing apparatus may be necessary.

Treatment of a flammable spill should not only absorb the material but also reduce the fire hazard. A procedure that effectively contains the liquid and reduces its vapor pressure is preferred. Water should not be used in cleaning up a flammable spill. For example, a 5% solution of alcohol in water still shows an appreciable vapor pressure. Again, commercial clay absorbent or kitty litter is the best approach. The absorbent should be poured first around the perimeter of the spill and then inward to cover the spill. Common flammable liquids in the laboratory are listed in Table 12-1. See Chap. 8 for detailed information on handling leaking cylinders of flammable gases.

Toxic Substances

Toxic substances require special safeguards. The general spill cleanup procedure involves the use of a reagent or materials to lessen or eliminate the toxic character of the spill prior to cleanup. Try to avoid skin contact and inhalation. Use the appropriate breathing apparatus. If any material spills or splashes on clothing, the garment should be immediately removed to prevent skin contact.

Consider, for example, the hazards of a 1-L bottle of toluene dropped in a small laboratory 30 ft × 10 ft containing 67 m^3 of air. A liter bottle contains about 866 g of toluene. The vapor pressure of toluene is such that a spill of that order would quickly vaporize at 25°C. In a lab of 67 m^3, the concentration would be 12,900 mg/m^3 or 3600 ppm, 36 times the TLV of 100 ppm. A definite health hazard would exist. Toluene is also a substantial fire hazard. Its flashpoint is 4.4°C. The lower explosive limit (LEL) is 1.3%, and the upper explosive limit (UEL) is 7.1%. The autoignition temperature is 536°C or 997°F. In a lab of our hypothetical model, the percentage of toluene in the air would be 0.36% toluene and no fire or explosion hazard would exist if the toluene were thoroughly mixed with air. However, in the immediate vicinity of the spill, the LEL could be exceeded.

Since the nature of toxic materials varies so greatly, refer to the National Research Council's *Prudent Practices for Disposal of Chemicals from Laboratories,* or the Manufacturing Chemists Association's *Laboratory Waste Disposal Manual* or *Guide for Safety in the Chemical Laboratory* for detailed treatment of a diverse array of chemicals.

Mercury. Mercury spills present special problems because of the high toxicity of mercury vapor. Mercury vapor is about 100 times more poisonous than hydrogen cyanide. It is also a cumulative poison. Spilled mercury should be immediately and thoroughly cleaned up using specialized cleaning equipment. If a mercury cleanup unit is available, become familiar with its location and proper use.

All spilled mercury must be collected and returned to a covered container. If a mercury spill kit is not available, brush the mercury carefully into a dustpan, sprinkle the affected area with sulfur, and clean again. A glass capillary connected to an aspirator can be used as a fine-tip vacuum cleaner to pick up the droplets. Mercury spilled into floor cracks can be made nonvolatile by amalgamation with zinc dust. Domestic vacuum cleaners must not be used because they will only disperse mercury aerosols and spread the contamination. A mercury-vapor-monitoring instrument

TABLE 12-1. Flammable Liquids Commonly Found in Laboratories

Compound	Bp (°F)	Ignition (°F)	TWA (ppm)	Flash-point	LEL (% vol.)	UEL (% vol.)
Acetone	134	869	1000	0	2.6	12.8
Acetylacetone	136	N/A	N/A	105	N/A	N/A
Carbon disulfide	115	257	20	−22	1.3	50.0
Ether	95	320	400	−49	1.9	36.0
Ethyl acetate	171	800	400	24	2.2	11.0
Ethyl alcohol	79	423	1000	55	3.3	19.0
Hexane	156	437	100	−7	1.1	7.5
Methyl alcohol	147	725	200	52	6.7	36.0
Methyl ethyl ketone	80	515	200	21	1.8	10.0
Methyl isobutyl ketone	117	460	900	73	1.4	7.5
Toluene	231	986	100	40	1.2	7.1

LEL: Lower explosive limit (the lowest concentration at which a flame will be propagated, leanest burnable mixture).
UEL: Upper explosive limit (the richest burning mixture).
Source: From NIOSH, *Hazards in the Chemistry Laboratory, Student Manual,* 1978, p. 3-5.

should be available for determining the effectiveness of the cleanup.

Even small amounts of mercury are dangerous. If a mercury thermometer is broken, approximately 1 g of mercury would be spilled. At room temperature (24°C), the vapor pressure of mercury is 0.001591 torr. The mercury will vaporize until an equilibrium concentration of 20 mg/m^3 is achieved. If the laboratory were 50 ft × 25 ft × 10 ft, it would contain about 330 m^3 of air. If poorly ventilated, the room could contain up to 6600 mg of mercury vapor or six times as much as the spill. Thus, it is safe to assume that all the mercury would then vaporize. At that rate, the air in the room would contain 3.03 mg/m^3 mercury vapor as compared to the TLV of 0.05 mg/m^3 allowed. Thus, the spillage of the mercury in one thermometer can pose a significant laboratory hazard.

Mercury is found in flow meters, thermometers, air pressure gauges, mercury diffusion pumps, mercury vapor lamps, and dental fillings. Laboratories which routinely work with mercury would benefit from preventive forethought. For example, floors should be smooth and free from cracks in which mercury can hide.

CHEMICAL AND PHYSICAL PROPERTIES OF MERCURY

1. High density.
2. High surface tension—tends to form spheres.
3. Low viscosity—encourages splash and the resulting small spheres will adhere to even vertical surfaces.
4. May form an oxide skin on exposure to air, which reduces evaporation. However, vibration could break the oxide layer.
5. Vaporizes readily at room temperature especially from small droplets (a 2.8-mg droplet if evaporated in a room 10 ft × 10 ft × 10 ft will generate 0.1 mg/m^3 of mercury vapor, twice the ACGIH's TWA. Mercury near heat sources (radiators, heat ducts, motors, ovens, and lab heaters) will evaporate faster.

SPILL CLEAN-UP

1. Resistant to cleaning solutions.
2. Insoluble in water, alkalies, common solvents, dilute sulfuric acid.
3. Dissolves in dilute nitric acid and hot, concentrated sulfuric acid.
4. Amalgamates with zinc dust.
5. Use a specialized glass vacuum system. Mercury may amalgamate with metals in normal vacuum systems.
6. Specialized mercury cleanup packages are available.

CONTROL MEASURES

1. Do not handle mercury in open room air. Use fume hoods and glove boxes. Do not handle mercury in high traffic areas.
2. When transferring mercury from one container to another, cover the mercury surface with oil or water. Use equipment and enclosures designed to localize splash and spills.

Corrosives

A corrosive chemical spill must be neutralized prior to other cleanup procedures. Corrosive chemicals are injurious to body tissue and corrosive to metals by direct action. Most are acids or alkalies and can be in the form of a pure solid, liquid, gas, or in solution. See Table 12-2 for a list of common corrosive chemicals.

Neutralization is usually accomplished by treatment with a mild acid or base. Acids, acid halides, and acid anhydrides are neutralized with sodium bicarbonate. Treat halogens with reducing agents such as sodium thiosulfate. Neutralize alkalies with dilute hydrochloric acid (3-6N) or a mild acid like boric acid. For acid chloride spills, just use a clay-type absorbent or dry sand.

Once neutralized, put an absorbent on the spill. Various commercial absorbents including cat litter can be used. Some commercial materials contain both a neutralizer and an absorber with an indicator. Some companies supply spill cleanup kits that include disposable agents, gloves, safety glasses, brushes, scoops, bags, and ties.

Anyone who cleans up corrosive spills should avoid any contact with the material by wearing breathing apparatus and other appropriate safety equipment. See Chap. 8 for detailed information on handling leaking cylinders of corrosive gases.

Oxidizing and Reducing Agents

Cleanup of strong oxidizing and reducing agents demands specialized procedures. Refer to the appropriate MSDS or the MCA's *Guide for Safety in the Chemical Laboratory.* Redox reactions can occur in the solid state, as well as the liquid or gas phase. *Never*

TABLE 12-2. Corrosive Chemicals in the Laboratory

Acids	Salts and Other Toxic Compounds
Acetic acid	Arsenic halides
Hydrochloric acid	Bromine
Hydrofluoric acid	Iodine
Nitric acid	Mercury(I) chloride
Perchloric acid	Mercury(II) chloride
Phosphoric acid	Phenol
Sulfuric acid	Potassium dichromate
	Potassium permanganate
Alkalies	***Gases***
Ammonium hydroxide	Chlorine
Calcium oxide	Hydrogen chloride
Potassium hydroxide	Hydrogen sulfide
Sodium carbonate	Nitrogen dioxide
Sodium hydroxide	Sulfur dioxide
Trisodium phosphate	

Source: From NIOSH, *Hazards in the Chemistry Laboratory, Student Manual,* 1978, p. 3–3.

attempt to clean up an oxidant by indiscriminately adding a reducing agent.

Alkali metal spills should be smothered with a graphite or Met-L-X extinguisher and removed to a safe location where they can be neutralized by reaction with a *dry* secondary alcohol. Particles of alkali metal splattered on the skin should be quickly removed and the skin flushed with water. If any metal particles on the skin ignite, flood with cold water immediately.

Sodium-potassium alloys (NaK) are very hazardous. Follow the suppliers' recommendations exactly.

A white (yellow) phosphorus spill should be covered with wet sand or a wet absorbent and disposed of by controlled burning outdoors. Be sure to consult local regulations on outdoor burning. Any white phosphorus in contact with the skin should be flushed off with cold water. Any remaining should be removed with a spatula or forceps. A copper sulfate solution can be used as an indicator to identify remaining particles. It reacts to form a dark color with elemental phosphorus.

Large Chemical Spills

For most small-scale laboratory spills, the procedures described in the preceding sections will be adequate. When large-scale spills may be possible, emergency procedures should be prepared in advance for containing spilled chemicals with minimal damage. A spill-control policy should include consideration of the following points:

1. *Prevention:* Storage, operating procedures, monitoring, inspection, and personnel training.
2. *Containment:* Engineering controls on storage facilities and equipment.
3. *Cleanup:* Countermeasures and training of designated personnel to help reduce the impact of a chemical spill.
4. *Reporting:* Provisions for reporting spills both internally (to identify controllable hazards) and externally (e.g., to state and federal regulatory agencies).

The primary objective of spill control is to minimize the hazard as far as possible. The first step in achieving this is to keep only minimal amounts of reagents in the laboratory. Plan carefully and purchase, at the most, a semester's supply of reagents, preferably less if possible. Do not purchase large quantities for a trivial saving of a few dollars at the expense of chemical safety. When chemical spills do occur, the following cleanup techniques should be implemented.

FIRES AND FIREFIGHTING

Fire has always been one of the attendant hazards of laboratory operation. Laboratories make frequent use of flammable materials including solids, liquids, gases, and vapors. The potential for extensive property damage and severe personnel injury is very high in the science laboratory. Assume fires will happen. The goal of every laboratory worker should be to reduce the chance of fire to the lowest probability possible.

Elements of a successful fire control program include adequate and effective education of individuals, both students and instructors, in the hazards of fire; instruction of personnel in the use of fire extinguishing equipment; the use of proper laboratory procedures; the maintenance of proper chemical storage facilities; and the provision and maintenance of effective fire control equipment.

Small laboratory fires may be controlled with equipment to be described later. However, large fires require professional help immediately. When in doubt, get assistance from the nearest fire department.

It is especially important to remember that chemical fires often result in new products that are very toxic. Ammonia, hydrogen cyanide, hydrogen chloride, carbon monoxide, nitrogen dioxide, sulfur dioxide, and phosgene accompany many fires fed by chemicals. The cleanup of a laboratory that has been damaged by fire may present special hazards due to the presence of these and other toxic chemicals in the atmosphere and work environment, and may require special precautions.

Fire Prevention

Fire is a common hazard in laboratory operations. Fire prevention is the first step toward safety. The following are some common approaches to fire prevention for the laboratory.

1. Purchase reagents in the smallest quantities possible.
2. Do not store incompatible reagents together (see *Academic Laboratory Chemical Hazards Guidebook*, Chap. 1 for a list). Compatibility is more important than alphabetizing the collection. Also in *Academic Laboratory Chemical Hazards Guidebook*, Chap. 2 and Appendix G provide information on chemical storage systems to minimize hazards.
3. Keep flammable liquids in appropriate steel safety cans.
4. Store flammable liquids unsuited to cans in special safety cabinets.
5. Do not store flammable liquids in standard refrigerators.

Facilities. Facilities should be designed to allow a safe exit if a major fire occurs. All exits should be marked and lighted, cleared and not blocked. The location of fire extinguishers should be clearly marked. Doors should have fire bars for quick action and all doors should be kept closed. Smoke detectors and fire control systems should be installed. Laboratories should be equipped with sprinkler systems, except where minimal chemicals are used and much electronic equipment is maintained.

Storage facilities should be constructed and maintained with the potential of fire in mind. Corrosives, flammables, toxic substances, and oxidizers must be stored in separate isolated areas to prevent mixing in case of fire. Remember, that in case of a fire, even a small container of an exploding flammable will scramble the materials in a storage area.

Work areas should be provided with appropriate hazard control equipment. No smoking should be allowed in flammable areas. Flammable liquids should be stored in metal safety cans that are grounded when transferring them. Only the smallest quantity of flammable liquids should be kept on hand. See Table 12-3.

Storage of Flammable Chemicals. Proper chemical storage is a critical factor in any science laboratory fire control program. There is one cardinal rule to be observed in acquiring and storing chemical compounds: Chemical reagents should be purchased and stored in the smallest quantities possible. Purchasing philosophies generally require that materials, from chemical reagents to nuts and bolts, be purchased in the most cost-effective manner. For chemical reagents, cost-effective is synonymous with large quantity and large packages. Cost should not be a factor in the acquisition of chemical reagents for the school science laboratory. Safety and health considerations are much more important.

It is recommended that a single quarter or at most a single semester's supply of a chemical be bought at one time. Most chemical reagents can be delivered by supply houses in 24–48 h in most areas of the country. Flammable solvents should be purchased in the smallest feasible containers. If the container is glass, it should have a protective plastic coating.

The storage of flammable compounds is a complex problem. Strict inventory should be kept and a reference file should be maintained listing flashpoint, boiling point, explosive limits, auto-ignition temperatures, products of combustion, and extinguishing agents. Large quantities of flammable liquids should be stored outside in an area protected from the elements. A limit of 100 drums 60 ft from the building is recommended by the MCA. Drains and fire hydrants should be provided. Inside storage should be as lim-

TABLE 12-3. OSHA Regulations for Storing Flammable Liquids

			Type of Container		
Class	**Flashpoint (°F)**	**Boiling Point (°F)**	**Plastic or Glass**	**Metal**	**Safety**
Flammable Class					
1A	FP<73	BP<100	1 pt	1 gal	2 gal
1B	FP<73	BP>100	1 qt	5 gal	5 gal
1C	73<FP<100		1 gal	5 gal	5 gal
Combustible Class					
II	100<FP<140		1 gal	5 gal	5 gal
III	FP>140		1 gal	5 gal	5 gal

Source: From NIOSH, *Hazards in the Chemistry Laboratory, Student Manual,* 1978, p. 3–11.

ited as possible. A flammable liquid storage vault, supplied with a sprinkler and appropriate ventilation, should be available. Cans of 5-gal containers should not be used. Maximum volumes of 1 L should be stored in a safe manner in the laboratory. Storage problems are compounded by the varied nature and sizes of containers.

Laboratory Flammables Storage Areas. In most educational institutions, both flammable and nonflammable chemical reagents will be stored in the same storage area. Therefore, the storage area should be designed in such a manner that it effectively deals with the key elements of chemical storage discussed in *Academic Laboratory Chemical Hazards Guidebook,* Chap. 4.

The National Fire Protection Association (NFPA) publishes standards that govern fire protection in educational institutions. The NFPA code states that the storage of flammable liquids shall be limited to that required for maintenance, demonstration, treatment, and laboratory work. The code establishes the following storage provisions for flammable liquids:

1. No container shall exceed a capacity of 1 gal.
2. Not more than 10 gal of flammable or combustible liquids shall be stored outside of a storage cabinet or storage room except in safety cans.
3. Not more than 25 gal of flammable or combustible liquids shall be stored in safety cans outside of a storage room or storage cabinet.
4. Quantities of flammable and combustible liquids in excess of those set forth in this section shall be stored in an inside storage room or storage cabinet.

Storage Rooms for Flammable Liquids. School science laboratories sometimes have a specially built room or special buildings for the storage of flammable chemicals. A special isolated storage room is recommended for laboratory facilities that stock over 50 gal of flammable liquid. This is strictly a recommendation and not a legal requirement.

OSHA regulations require that inside storage rooms have at least one clear aisle with a minimum width of 3 ft. The exit(s) for the inside storage room must be clearly marked and not blocked in any way. OSHA regulations or NFPA 30 should be consulted for other important design factors.

For purposes of fire protection, when flammable liquids are stored, the NFPA code requires that:

1. The floor in the storage room be at least 4 in. lower than the floors of the surrounding rooms and corridors, or that there be a 4-in. high sill between the storage area and adjacent areas.
2. All doors be approved, self-closing fire doors. [See NFPA 30.4310 and 29 CFR, 1910.106(d)(4)(i).]
3. The room be liquid-tight where the walls join the floors.

The code also specifies that storage rooms with a maximum floor area of 150 ft^2 must have walls, floor, and ceiling with a fire resistive rating of at least 1 h. Larger storerooms must have a fire-resistive rating of at least 2 h.

The quantities of flammable liquids stored are limited by the code to 2 gal/ft^2 of floor area for unprotected storage rooms less than 150 ft^2 in area. If the storage room has a fire protection system, however, the quantity stored can be increased to 5 gal/ft^2. A fire protection system is a sprinkler system, a carbon dioxide foam system, or any similar automatic system that is acceptable to the local authorities.

The floors in storage rooms should be constructed of a material that possesses good chemical resistance and is readily cleaned. All electrical outlets and equipment must be well grounded.

If there are no special, central storage areas for chemical storage, and flammable liquids are stored or used in a school science laboratory, the laboratory must be separated from nonlaboratory areas by construction having at least a 1-h fire resistance.

Ventilation and Environment Control. All inside storage rooms should be equipped with a gravity exhaust system or a mechanical exhaust system to remove hazardous vapors. The exhaust duct should be located within 12 in. of the floor level. The supply air duct should be located on the opposite wall in a position that will minimize short-circuiting of the airflow pattern.

Powered mechanical exhaust systems must be provided for storage rooms in which liquids with flashpoints below 100°F are stored. The NFPA code requires that the exhaust system be capable of at least six changes of room air per hour. A further recommendation is that the exhaust rate should be 1 ft^3 per minute of exhaust per square foot of floor area, but not less than 150 ft^3/min. The ventilation system should also be capable of removing 10,000 ft^3 of air for every gallon of liquid vaporized. The room should be kept cool but not cold.

Drum Storage. Fifty-five gal drums are commonly used to ship flammable liquids, but are not intended as long-time inside storage containers. It is not safe to dispense from sealed drums exactly as they are received. The bung should be removed and replaced by an approved pressure and vacuum relief vent to protect against internal pressure buildup in the event of fire or if the drum might be exposed to direct sunlight.

If possible, drums should be stored on metal racks placed such that the end bung openings are toward an aisle and the side bung openings are on top. The drums, as well as the racks, should be grounded with a minimum length of American wire gauge 10 wire. Because effective grounding requires metal-to-metal contact, all dirt, paint, and corrosion must be removed from the contact areas. Spring-type battery clamps and a minimally sized conductor (e.g., American wire gauge 8 or 10) are satisfactory. It is also necessary to provide bonding to metal receiving containers to prevent accumulation of static electricity (which will discharge to the ground, creating a spark that could ignite the flammable vapors). Drip pans that have flame arresters should be installed or placed under faucets.

Dispensing from drums is usually accomplished by one of two methods. The first is gravity based through drum faucets that are self-closing and require constant hand pressure for operation. Faucets of plastic construction are not generally acceptable due to chemical action on the plastic materials.

The second, and safer, method is to use an approved hand-operated rotary transfer pump. Such pumps have metering options and permit immediate cutoff control to prevent overflow and spillage, can be reversed to siphon off excess liquid in case of overfilling, and can be equipped with drip returns so that any excess liquid can be returned to the drum.

Safety Cans. Many research laboratories use "safety cans" for storing flammable liquids, both reagents and wastes. Safety cans are stainless steel or coated steel cans designed to minimize the probability of ignition of flammable vapors and avoid the accidental breakage of a flammable liquid container, usually glass, which may occur in the typical science laboratory.

Safety cans are equipped with spring-loaded closures and have flame arrestors in the spout. The flame arrestor consists of a baffle screen that smothers any flame before it can enter the can.

Safety cans do have certain characteristics that tend to inhibit their use in some schools. They are costly, and they generally cannot be used to store high-purity flammable liquids. In many cases, purity should not be a critical factor in school science laboratories. Cost may be a factor, but the potential fire hazard presented by flammable liquids should justify the relatively small additional cost of safety cans.

Safety cans used for flammables waste disposal present other problems. People get tired of opening the lid each time they pour something in and eventually will use some means of holding the top open. Although the flame arrestor would be effective in preventing a flash into the can, the open lid allows the spread of vapors into the room, and if they are flammable or toxic, some hazard exists. These cans should be kept closed at all times, except when material is being put into them or they are being emptied. Another problem is that although they will last a long time with hydrocarbons, water-miscible solvents such as alcohols can cause severe corrosion. Chlorinated hydrocarbons decompose to yield hydrogen chloride that is also very corrosive to metal and can quickly ruin an expensive disposal can.

Safety cans are carefully designed to provide protection against fire. They should not be modified in any fashion for the purpose of increasing the filling or dispensing rate of the can.

Flammable Liquid Cabinets. Flammable liquid cabinets are often used in science laboratories and storerooms to provide both protection against fire and security against improper chemical usage. Safety cabinets can be made of double-walled steel construction or wood. They are equipped with locks to secure the contents and with plumbing connections that permit the connection of the cabinets to a forced-air ventilation system.

The NFPA has developed regulations for the use of both wooden and steel cabinets. According to the NFPA, cabinets should be designed so that their internal temperature does not exceed 325°F when subjected to a 10-min fire test using the standard time-temperature curve specified by the NFPA. The code requires that the bottom, top, door, and sides of steel cabinets be double-walled with at least a 1.5 in. air space between the walls.

Tests made by the Los Angeles Fire Department have shown properly constructed wooden cabinets to be at least as effective as, and in many cases better than, the steel cabinets. NFPA 30 and the OSHA regulations contain specifications for the construction of wooden cabinets. The NFPA specifies that all wood cabinets used for the storage of flammable liquids be constructed of wood at least 1 in. thick. The Los Angeles Fire Department specifies that wood cabinets used for the storage of "dangerous" chemicals be constructed of wood at least 2 in. thick, and wood cabinets used for the storage of flammable chemicals be at least 1 in. thick.

The NFPA code specifies that not more than 60 gal of flammable or 120 gal of combustible liquids may be stored in a storage cabinet.

Flammable storage cabinets provide a convenient method for storing flammable and toxic chemicals when a central storage facility is not available. Those equipped with a lock provide security in school situations. When connected to a mechanical exhaust system, they remove any reason to store chemicals in a fume hood. To prevent the accumulation of toxic or explosive chemical vapors, one manufacturer of flammable storage cabinets recommends that they be exhausted at a rate of 20 ft^3/min (cfm).

Explosion-Proof Refrigerators. Refrigerators are common accessories in school science laboratories. Refrigerators are used to store biological materials, as well as highly volatile chemical reagents. Highly volatile organic solvents must not be stored in standard domestic refrigerators. Many refrigerators have exploded when flammable vapors were released and ignited by a sparking thermostat.

Refrigerators for flammable liquid storage in the laboratory are available in two special designs: "explosion-proof" and "explosion-safe." In explosion-proof refrigerators, the electrical components are enclosed in explosion-proof housings both inside and out. Hence, they may be used to store flammables and also can be used when flammable vapors are present on the outside of the unit. Explosion-safe models have no ignition sources on the inside, but are not suitable for use in areas where flammable vapors can be present on the outside. The latter is less expensive and is usually adequate for a school laboratory.

Other Ways to Reduce Fire Hazards. Fire can easily occur in any laboratory. An understanding of the nature of fire is essential if a proper and adequate fire control program is to be established. In order for a fire to occur, all of the necessary components that make up the fire reaction must present. There are three components necessary for the vapor-phase reaction called fire. These are:

1. A supply of fuel
2. A source of heat or ignition
3. A source of oxygen

If any of these three components is absent, a fire cannot start or continue.

Reducing the presence of the controllable components in the laboratory will go a long way toward minimizing the chance of fire. The control of some fuels, flammable liquids, has already been discussed. The presence of oxygen is, of course, one factor that generally cannot be controlled. All laboratory workers should take great pains to control heat and ignition sources in the laboratory.

Ignition Sources. Gas burners commonly used in laboratories are indispensable for many purposes. Occasionally, they are used improperly. Flame burners must be treated as carefully as any other source of open flame and should be extinguished immediately if flammables are spilled or released. They should *never* be used as a source of heat for evaporating flammable liquids. Hot water baths and steam baths can often be used instead of burners when flammables are involved.

Whenever possible, electric mantles and hot plates should be used whenever a source of heat is required in the laboratory. Electric heaters do not present the fire hazard or danger to personnel that an open flame does.

Proper Clothing. Both students and instructors should wear appropriate clothing in the science laboratory. When selecting clothing for laboratory wear, prime consideration is given to protecting the wearer against chemical contact. However, the ability of the clothing to protect the wearer against fire must also be considered. Fire is an ever-present danger. Certain clothing materials are much greater fire hazards than others. Polyester fabrics will burn much more readily than cotton fabric will.

Students should be encouraged to wear cotton clothing whenever possible in the laboratory. Cotton lab coats provide some protection against fire. Many of the disposable lab coat fabrics burn quite readily despite the claims of the manufacturer. Any laboratory using these coats should thoroughly investigate their fire-resistant properties before purchasing them.

See Chap. 10 for recommendations on clothing to be worn in a laboratory.

Firefighting

Small Fires. Small fires that can easily be extinguished without evacuating the building or calling the fire department are among the most common laboratory incidents. However, even a minor fire can quickly become a serious problem. The first few minutes after discovery of a fire are critical in preventing a larger emergency. The following actions should be taken by laboratory personnel in case of a minor fire:

1. Attack the fire immediately, but never attempt to fight a fire alone. A fire in a small vessel can often be suffocated by covering the vessel with an inverted beaker or a watch glass. Do not use dry towels or cloths. Remove nearby flammable materials to avoid the possible spread of the fire.
2. In the event of fires that appear to be controllable, use the proper extinguisher. Direct the discharge from a 5- or 10-lb carbon dioxide, dry-chemical, or halon fire extinguisher at the base of the flames. Start at one side and work across the base. Always fight a fire from a position of escape.
3. If a spilled or sprayed liquid is burning over an area too large for the fire to be suffocated quickly and simply, all persons should leave except the instructor and those designated to help—for example, a particular individual to call the fire department and sound the necessary alarms.
4. Avoid being trapped by the fire; always fight a fire from a position accessible to an exit.
5. Avoid breathing fumes from the fire. Toxic gases and smoke may be present. If necessary, abandon the fire until self-contained breathing apparatus can be used.
6. Fires involving very reactive metals should only be fought with Class D fire extinguishers, powdered graphite, or Pyrene G-1. Carbon dioxide and the usual dry-chemical fire extinguishers will aggravate fires of many metals such as alkalis, alkaline earths, aluminum, magnesium, zirconium, hafnium, thorium, and uranium.
7. Never forget the possibility of explosion. Special care should be taken to keep fire or excessive heat from volatile solvents, compressed gas cylinders, reactive metals, and explosive compounds.
8. Immediately after the fire, all extinguishers that were used should be refilled or replaced with full ones.

Large Fires. If there is any doubt whether the fire can be controlled by locally available personnel and equipment, the following actions should be taken:

1. Notify the fire department and activate the emergency alarm system.
2. Confine the emergency. Close hood sashes, doors between laboratories, and fire doors to prevent the further spread of the fire.
3. Assist injured personnel. Provide first aid or transportation to medical aid if necessary.
4. Evacuate the building to avoid further danger to personnel.
5. When they arrive, firefighters should be informed of what chemicals are involved, or which chemicals may become involved. A current inventory list is very helpful.

Explosions. Explosions occur when flammables are thoroughly mixed with oxidants and the ensuing reaction is rapid and violent. Victims of explosions are often subject to shock and must be treated for this condition.

In case of an explosion, immediately turn off burners and other heating devices, stop any reactions in progress, assist in treating victims, and vacate the area until it has been decontaminated.

It is the responsibility of the laboratory supervisor to determine whether unusual hazards exist that require more stringent safety precautions. In large laboratories, or when risk is high, designated firefighting teams may be necessary to minimize risk. Special arrangements with local fire departments to warn them of the hazards of chemical fires may be desirable in some situations.

Classes of Fires and Fire Suppression Equipment

There are four recognized classes of fires: Class A, Class B, Class C, and Class D.

Class A fires: Those that occur in ordinary combustible materials including wood, paper, cloth, coal, rubber, textiles, and plastics.

Class B fires: Those that are fueled by flammable liquids—gasoline, motor oil, grease, mineral spirits, alcohol, ether, etc.

Class C fires: Those that originate in electrical equipment.

Class D fires: Those that are fueled by combustible metals (e.g., sodium, potassium, and magnesium), metal hydrides, or organometallics (such as alkylaluminums).

Fire extinguishers are labeled with an A, B, C, or D or any combination of these designations to indicate which classes of fires they can be used to extinguish. The labels on extinguishers also contain directions for their use. The instructor and students should be familiar with the operating instructions for all fire extinguishers in the laboratory.

Class A fires can be extinguished with water, dry chemical, or halogenated hydrocarbon (halon) portable extinguishers. The ABC dry chemical unit is extremely effective and is preferred. Water may also be quite adequate and has the advantage of not leaving a powder residue over the area.

Class B fires can be extinguished by carbon dioxide (CO_2), dry chemical, or halon extinguishers. Carbon dioxide extinguishers are excellent for small flammable liquid fires when properly used. Care must be used with CO_2 extinguishers so that the blast from the extinguisher does not spread the fire. For larger Class B fires, either BC or ABC dry chemical extinguishers are more effective.

Class C fires, in electrical equipment, can be extinguished by carbon dioxide, halon, or dry chemical extinguishers **after the current has been shut off.** Water should be avoided because it is a conductor.

Class D fires require special extinguishing agents applied from an extinguisher or shoveled from a bucket. Sodium and other such metals may react with carbon dioxide or halon.

In case a fire breaks out in the lab, be sure you understand the proper use and application of firefighting equipment. The right extinguisher must be used in fighting a fire.

Fire Extinguishers. Successful fire control in the school laboratory is not just a matter of having a fire extinguisher, but having the right kind of fire extinguisher and someone who knows how to use it.

The proper location of extinguishers is critical. They must be near enough to procure and use without delay, generally 50 ft or less. They should all be located at room exits, not deep in the room. When a person goes for an extinguisher, he or she should be going toward safe egress.

All chemical laboratories should be provided with carbon dioxide or dry chemical fire extinguishers (or both). Other types of extinguishers should be available if required by the work in progress. The four types of extinguishers most commonly used are classified by the type of fire (see above) for which they are suitable. Labels on fire extinguishers contain directions for their use. All personnel should be trained in the use of extinguishers for each class of fire since every source is present in an analytical laboratory.

OSHA requires the monthly inspections of fire extinguishers in commercial and government establishments. Local fire regulations usually apply to schools.

Soda Acid Extinguishers. Soda acid (water) extinguishers are effective against burning paper and trash (Class A fires). These should not be used for extinguishing electrical, liquid, or metal fires.

Carbon Dioxide (CO_2) Extinguishers. Carbon dioxide extinguishers are effective against burning liquids, such as hydrocarbons or paint, and electrical fires (Class B and C fires). They are recommended for fires involving delicate instruments and optical systems because they do not damage such equipment. They are less effective against paper and trash or metal fires and should not be used against Class D fires, especially lithium aluminum hydride fires.

Be careful when using a CO_2 extinguisher. The blast from the nozzle can spread a liquid fire or knock bottles over or off shelving.

Dry Chemical Extinguishers. Normal dry powder extinguishers, which contain sodium bicarbonate or monoammonium phosphate, are effective against burning trash, liquids, and electrical fires (Class A, B, and C fires). They are less effective against metal fires. They are not recommended for fires involving delicate instruments or optical systems because of the cleanup problems created. These extinguishers are generally used when large quantities of solvent may be present.

Met-L-X[RT] extinguishers and others with special granular formulations are effective against burning metal (Class D fires). Included in this category are fires involving magnesium, lithium, sodium, and potassium; alloys of reactive metals; and metal hydrides, metal alkyls, and other organometallics. These extinguishers are less effective against paper and trash, liquid, or electrical fires.

Halon Extinguishers. Because a fire in an electrical cabinet can severely damage expensive components and can force extended project delays, an efficient, fast-acting extinguishing system is necessary. Water sprinkler systems are not acceptable in this situation, as they could cause as much damage to the electrical components as the fire and smoke itself.

Halons are colorless, odorless, electrically nonconductive gases that extinguish fires by inhibiting the chemical reaction of fuel and oxygen. Halon was developed in the late 1940s by the U.S. Army, which was searching for an extinguishing agent with good firefighting properties but without the high toxicity of the other materials on hand at the time. Strong points for choosing halon over other types of extinguishing chemicals include its effectiveness controlling Class A (cellulosic materials), Class B (flammable liquids), and Class C (electrical) fires. Halon is ideal for extinguishing fires in electrical cabinets because it is a gas and can penetrate even difficult places. Also, a concentration of only about 5% by volume is enough to extinguish most fires. After discharge, it leaves no residue on electrical components.

Every extinguisher should carry a label indicating what class or classes of fires it is effective against. There are a number of other more specialized types of extinguishers available for unusual fire hazard situations. Each laboratory worker should be responsible for knowing the location, operation, and limitations of the fire extinguishers in his or her work area. It is the responsibility of the laboratory supervisor to ensure that all laboratory workers are shown the locations of nearby fire extinguishers and are trained in their use. After use, an extinguisher should be recharged or replaced by designated personnel.

Fire Extinguisher Records. Fire extinguishers must be readily available in school science laboratories. When needed, they must work. In order to ensure that an extinguisher will function when needed, a program of regular inspection, maintenance, and repair must be established and operated.

Fire extinguisher records should list extinguishers by type, location, recharge periods, and size. Records of available spare parts and spare extinguishers must also be maintained. Generally, fire extinguisher maintenance will be performed by an outside service company. Their visits should be scheduled in advance and noted in a written record.

All pressurized fire extinguishers must be monitored to ensure that the pressure is sufficient to provide the necessary propulsive force for the contents of the extinguisher. Nearly all commercial extinguishers have pressure gauges indicating the current status of the extinguisher charge. Most extinguisher users contract out the inspection and recharging of extinguishers. Inspections of extinguishers are made on a regular basis by the contractor, and those that require recharging are recharged.

Recharging can be done on-site or at a remote location. If the extinguishers are removed from the school or laboratory area, a reserve supply of extinguishers may be necessary to maintain protection at the required level.

OSHA regulations require that extinguishers be checked monthly to see that they are in place, their seals are unbroken, and they are accessible.

Fire Hoses. Fire hoses are intended for use by trained firefighting personnel against fires too large to be handled by extinguishers, but they are included as safety equipment in some structures. Water has a cooling action and is effective against fires involving paper, wood, rags, trash, etc. (Class A fires). Water should not be used directly on fires involving live electrical equipment (Class C fires) or chemicals such as alkali metals, metal hydrides, and metal alkyls that react vigorously with it (Class D fires).

Streams of water should not be used against fires that involve oils or other water-insoluble flammable liquids (Class B fires). This form of water will not readily extinguish such fires, and it will usually spread or float the fire to adjacent areas. These possibilities are minimized by the use of a water fog.

Water fogs are used extensively by the petroleum industry because of their fire-controlling and extinguishing properties. A fog can be used safely and effectively against fires that involve oil products, as well as those involving wood, rags, rubbish, etc.

Because of the potential hazards in using water around chemicals, laboratory workers should refrain from using fire hoses except in extreme emergencies. A hose delivering water at significant pressures is difficult to control and can knock chemical containers over or off shelves. Such use should be reserved for trained firefighting personnel.

Automatic Fire-Extinguishing Systems. In areas where fire potential (e.g., solvent storage areas) and the risk of injury or damage are high, automatic fire-extinguishing systems are often used. These may be of the water-sprinkler, carbon dioxide, dry chemical, or halogenated hydrocarbon types. Whenever it has been determined that the risk justifies an automatic fire-extinguishing system, laboratory workers should be informed of its presence and advised of any safety precautions required for its action (e.g., evacuation before a carbon dioxide total-flood system is actuated).

Quick On/Off Sprinklers. At many areas, fire-suppression water must be minimized, such as in areas containing expensive water-sensitive equipment (e.g., computer rooms and experiment control rooms). A solution to this problem may be quick-acting (QA) on/off sprinkler heads, which are designed to respond quickly and shut off automatically when a fire is controlled or extinguished. The QA sprinkler has a two-phase operational mechanism: an eutectic material holds the deflector housing to prevent water leakage and a snap disk (to the side) that activates the on/off action. The eutectic holding the deflector is designed to melt at 57°C, which clears the path for water flow; the snap disk is set to release water at 74°C. The ordinary on/off sprinkler requires only the operation of the snap disk, which is also set at 74°C.

When these sprinkler heads operate successfully, they extinguish a fire with approximately 55–90% less water than that required by conventional heads. The effectiveness of these sprinkler heads is primarily due to quick response and a dense spray pattern. However, one potential problem with this fast response is the short activation time for each cycle, typically 10–20 sec. In most cases, the cycle time is sufficient to control the fire until someone arrives to extinguish it manually; however, the QA sprinkler may apply the water for too brief a time to automatically put out certain fires. The total quantity of water discharged by a conventional sprinkler is a function of how soon personnel can determine that the fire is out, locate the shut-off valve, and deactivate the sprinklers. QA on/off sprinklers should be considered for fire control in water-sensitive areas.

Fire Blankets. Many laboratories have fire blankets available. A fire blanket should be used primarily as a first-aid measure for the prevention of shock rather than against smoldering or burning clothing. It should be used only as a last-resort measure to extinguish clothing fires; such blankets tend to hold heat in and increase the severity of burns. Clothing fires should be extinguished by immediately using a safety shower. If no shower is available, the victim should roll on the floor until the flame is out.

Student Education

Student education in fire control and fire protection in the laboratory is essential in the operation of any school science program. Students must be educated in all aspects of fire control, as well as the normal fire dangers inherent in laboratory operations.

Teachers should have hands-on training in the proper use of various fire extinguishers. The local fire department is usually more than happy to arrange a series of demonstration fires and give practice in extinguishment.

SUGGESTED READINGS

American Chemical Society, *Safety in Academic Chemistry Laboratories,* Washington, D.C., 1985.

Chemical Manufacturers' Association, *Laboratory Waste Disposal Manual,* Washington, D.C., 1972 and 1975 editions.

Manufacturing Chemists Association, *Guide for Safety in the Chemical Laboratory,* Van Nostrand Reinhold, New York, 1972.

National Fire Protection Association, *Fire Protection Guide on Hazardous Materials,* 9th ed., Quincy, Mass., 1986.

National Research Council, *Prudent Practices for Disposal of Chemicals from Laboratories,* National Academy Press, Washington, D.C., 1983.

13
FIRST AID

Medical emergencies that should be planned for are:

- Poisoning by ingestion, inhalation, skin absorption, or injection
- Chemicals (including liquids, dust, or glass) in the eye
- Asphyxiation (chemical or electrical)
- Wounds, especially if chemicals are present
- Burns, both from chemicals or heat
- Skin irritation by chemicals
- Lachrymatory vapor irritations

WHAT IS FIRST AID?

First aid is the immediate care of a person who has been injured or has suddenly taken ill. It is intended to prevent death or further illness and injury and to relieve pain until medical aid can be obtained. The overall objectives of first aid are (1) to control conditions that might endanger life; (2) to prevent further injury; (3) to relieve pain, prevent contamination, and treat for shock; and (4) to make the patient as comfortable as possible.

The initial responsibility for first aid rests with the first person(s) at the scene, who should react quickly but in a calm and reassuring manner. The person assuming responsibility should immediately summon medical help (be explicit in reporting suspected types of injury or illness and requesting assistance). The injured person should not be moved except when necessary to prevent further injury. Laboratory workers should be encouraged to obtain training in first aid and cardiopulmonary resuscitation (CPR).

The first objective is to save life by:

1. Ensuring an open airway and maintaining breathing.
2. Preventing heavy loss of blood.
3. Giving first aid for poisoning.
4. Preventing or reducing shock.
5. Preventing further injury.
6. Sending for medical aid.

The first-aider should also:

1. Avoid panic.
2. Inspire confidence.
3. Do no more than necessary until professional help arrives.

Information provided here on first aid is not designed to supplant a complete course of instruction under the direction of a qualified instructor. Such instruction is strongly recommended for all people who work in the laboratory and especially personnel who must respond to hazardous materials spills. The American Red Cross provides a 1-day course that covers the essentials of CPR and standard first aid. Videotapes on CPR are widely available and are helpful refreshers for people who have already had an approved CPR course. The classes of injury and the appropriate first-aid responses are detailed below.

BREATHING STOPPED

A person who has stopped breathing will die if breathing is not restored immediately. If breathing is restored, victims who have stopped breathing need hospitalization. The following are major factors in breathing stoppage.

POISONOUS GASES IN THE AIR OR LACK OF OXYGEN

1. Move victim to fresh air.
2. Begin mouth-to-mouth breathing.

3. Control the source of poisonous gases, if possible.
4. Keep others away from the area.
5. *Do not* enter an enclosed area to rescue an unconscious victim without first being equipped with a self-contained or air-supplied breathing apparatus.

ELECTRIC SHOCK

1. If an electrical hazard persists indoors, open the main electrical breaker if an appropriate individual breaker cannot be immediately identified. If such a hazard exists outdoors, contact the power company to turn off the current.
2. *Do not touch* the victim until he or she is separated from the current.
3. Begin mouth-to-mouth resuscitation or cardiopulmonary resuscitation, if needed and if trained in this technique, as soon as the victim is free of contact with the current.
4. *Do not* try to remove a person from an out-of-doors wire unless you have had special training in this type of rescue work.

HEART ATTACK
LARYNGEAL OBSTRUCTION
ACCIDENT OR DROWNING

When breathing movements stop or the lips, tongue, and fingernails become blue, the need for help with breathing exists. When in doubt, begin artificial respiration. No harm can result from its use. Delay may cost the victim his or her life.

Artificial Respiration

General.

1. Seconds count. Start immediately.
2. Remove any obvious obstruction from the mouth and throat.
3. Place the victim in an appropriate position and begin artificial respiration.
4. Maintain a steady rhythm of 12 breaths per minute.
5. Maintain an open airway and periodically check the victim. Be ready to resume artificial respiration if necessary.
6. Call a physician.
7. *Do not* move the victim unless absolutely necessary in order to remove him or her from danger.
8. *Do not* wait or look for help.
9. *Do not* stop to loosen clothing or warm the victim.
10. *Do not give up.*

Mouth-to-Mouth Breathing for Adults.

1. Place the patient flat on his or her back on the floor and kneel at that person's side.
2. Establish an airway. Check the patient's mouth with your finger to be sure that no obstruction is present and then tip the patient's head back until his or her chin points straight up.
3. Pinch the victim's nose shut, with the thumb and forefinger of the hand on the victim's forehead. Begin mouth-to-mouth resuscitation by taking a deep breath and placing your mouth over the patient's mouth so as to make a leakproof seal. Blow your breath into the patient's mouth until you see his or her chest rise.
4. Remove your mouth and allow the patient to exhale.
5. Repeat the procedure at a rate of once every 5 sec.
6. Maintain the head tilt and again check the victim for breathing for approximately 5 sec.
7. Let the victim exhale while you take another deep breath. As soon as you hear the victim breathe out, replace your mouth over his or her mouth or nose and repeat the procedure.
8. Repeat this procedure of giving one breath, turning to look, listen, and feel for a return of air, and blowing again once every 5 sec (12 times per minute).

Because of the fear of communicable diseases, especially AIDS, many people are fearful of performing mouth-to-mouth respiration. Various devices called mouth-to-mask resuscitators are available to eliminate the risk and fear of disease (Fig. 13-1). These devices consist of a mask to fit over the victim's nose and mouth, a filter to prevent the transmission of bacteria and viruses, and a mouthpiece to be used by the rescuer. Special vents are used to direct the exhaled air away from the rescuer.

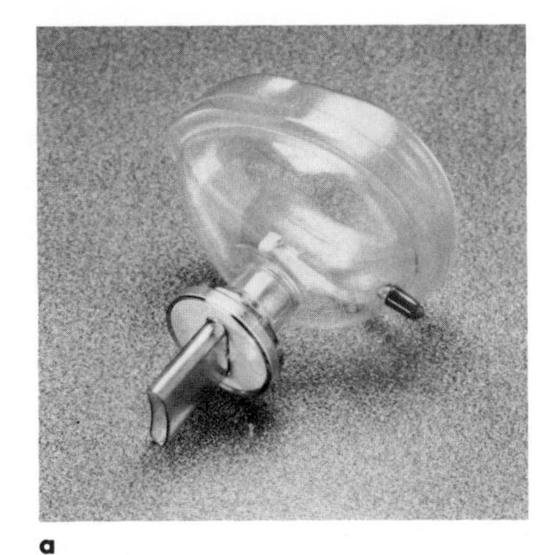

a

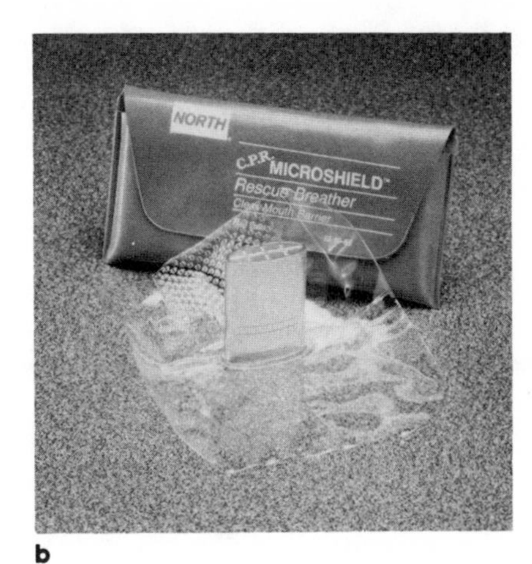

b

FIGURE 13-1. (a) Mouth-to-mask resuscitator and (b) disposable CPR microshield. (Photos courtesy of Lab Safety Supply, Inc., Janesville, Wis.)

Manual Method of Artificial Respiration. This method should be used when mouth-to-mouth resuscitation is advised against and a mouth-to-mask resuscitator is not available.

1. Place the victim in a face-up position, but allow his or her head to turn to the side to avoid aspiration.
2. Place something under the victim's shoulders to raise them to allow the head to drop backward.
3. Kneel above the victim's head, facing the victim.
4. Grasp the victim's arms at the wrists, crossing the wrists against the victim's lower chest.
5. Immediately, pull the arms upward, outward, and backward as far as possible.
6. Repeat 15 times per minute.
7. If a second person is present, he or she should hold the victim's head so that it tilts backward and the jaw juts forward.

Cardiopulmonary Resuscitation (CPR)

Heart-lung resuscitation is an emergency procedure that requires the ability to recognize a cardiac arrest and special training in its performance. Emergency cardiopulmonary resuscitation involves the following steps:

1. Airway opened
2. Breathing restored
3. Circulation restored

External cardiac compression should be started after providing four quick breaths and checking for pulse and breathing. If apnea (breathing stoppage) persists and unconsciousness, a death-like appearance, and the absence of a carotid pulse result, external cardiac compression should be started.

External cardiac compression consists of the application of rhythmic pressure over the lower half of the sternum. This compresses the heart and produces artificial circulation because the heart lies almost in the middle of the chest between the lower sternum and the spine.

External cardiac compression should always be accompanied by artificial respiration. To be effective, it requires sufficient pressure to depress the adult victim's lower sternum 1 1/2–2 in. (3.8–5.1 cm); the rate should be once a second. Considerably less effort will be required to achieve such depression in a child.

1. Check for a cardiac pulse; locate the larynx or adam's apple with the tips of the fingers and slide them into the groove between it and the muscle at the side of the neck. If no pulse is felt, circulation must be reestablished within 4 min to prevent brain damage.
2. The victim should be on his or her back on a firm surface. Kneel at the side of the patient near the waist and facing the patient's head.
3. Place the heel of your right hand over the heel of your left hand on top of the patient's breastbone about 1 in. (3 cm) above its lower tip.
4. Rock forward so that your shoulders are almost directly above the patient's chest. Keeping your arms straight and elbows locked, press almost vertically downward. Compress the patient's chest at least 1 1/2 in. (4 cm), then remove the pressure.

5. Continue at a rate of 60 times per minute. If you are working by yourself, that is, there is only one rescuer, perform both artificial circulation and artificial respiration using a 15:2 ratio, with two quick lung inflations after every 15 chest compressions. When two rescuers are present, optimum ventilation and circulation are achieved by quickly interposing one inflation after five chest compressions without any pause in compressions (5:1 ratio).
6. When two rescuers are available, one performs external cardiac compression, while the other keeps the patient's head tilted back and continues ventilation. Periodic palpation of the carotid pulse should be employed to check the effectiveness of external cardiac compressions or the return of a spontaneous heartbeat.

The preferred rate of 60 per minute is usually rapid enough to maintain blood flow and slow enough to allow cardiac refill. The compressions should be regular, smooth, and uninterrupted, with compression and relaxation being of equal duration. Under no circumstances should compression be interrupted for more than 5 sec. Every interruption in cardiac compression results in a drop of the blood pressure to zero.

Complications occurring from the use of cardiopulmonary resuscitation may include fracture of the ribs and sternum, laceration of the liver, and fat embolism. Several rules to follow are:

1. Never compress over the xiphoid process, the lower tip of the sternum. It extends down over the abdomen, and pressure on it may cause a dangerous laceration of the liver.
2. Never let the fingers touch the patient's ribs when compressing. Keep just the heel of the hand in the middle of the victim's chest over the lower half of his or her sternum.
3. Never use sudden or jerking movements to compress the chest.
4. Never compress the chest and abdomen simultaneously. This traps the liver and may cause it to rupture.

HEAVY BLEEDING

Heavy bleeding is caused by injury to one or more large blood vessels. A victim suffering from profuse bleeding may die within 1 min or less; therefore,

1. *Do not waste time.*
2. *Use pressure directly over the wound.*
3. Lay the patient down.
4. Place a pad, clean handkerchief, clean cloth, etc. directly over the wound and press firmly with one or both of your hands. If a pad or bandage is not available, close the wound with your hand or fingers.
5. Hold the pad firmly in place with a strong bandage, necktie, strips of cloth, etc. Unless bones are broken, raise the bleeding part higher than the rest of the body.
6. Keep the victim lying down.
7. Keep the victim warm to prevent loss of body heat. Cover with blankets, coat, or anything available and put something under him if he is on a cold or damp surface. *Do not* add heat.
8. Give fluids only if the victim does not have head or abdominal injuries, probably will not require surgery, and professional help will be more than 1 h in arriving. If the victim is conscious and can swallow, give her plenty of liquids to drink. Give her sips. Do not administer stimulants.
9. Call a physician.
10. Use a tourniquet only in cases of an amputation, or if the victim is bleeding profusely and other methods have failed and the victim's life is in danger. If a tourniquet is used, a record of the time it was applied must be kept.
11. *Do not* give the victim alcoholic drinks.
12. If the victim is *unconscious* or if abdominal injury is suspected, *do not* give him or her fluids.

CHEMICAL INGESTION

Before medical aid is available, the following should be done:

1. *Speed is essential.* Act before the victim's body has time to absorb the poison.
2. If the victim is conscious, give him or her water or milk immediately.
3. Call for medical aid immediately.
4. Attempt to learn exactly what substances were ingested and inform the medical staff (while the victim is en route to a hospital, if possible) and the local poison control center. The nature of the poison will determine the correct first-aid measure. First-aid information on labels may be incorrect; a physician or poison control center

should be able to provide proper advice on treatment.

5. Begin mouth-to-mouth resuscitation if the victim has difficulty breathing.
6. Decide if it is safe to induce vomiting.

Do not induce vomiting under the following circumstances (except on the advice of a doctor or poison control center):

1. If the victim is unconscious.
2. Is in convulsions.
3. Is known to have swallowed a petroleum product (kerosene, gasoline, lighter fluid), toilet bowl cleaner, rust remover, drain cleaner, lye, acids of a personal or household use, iodine, styptic pencil, washing soda, ammonia water, or household bleach, or has symptoms of severe pain, or a burning sensation in the mouth or throat.
4. If "do not induce vomiting" is indicated in the first-aid segment of the chemical data sheets.

To induce vomiting:

1. Induce vomiting by using 10 g of salt in 200 mL of warm water (2 teaspoonfuls in a glass of warm water) or use 30 mL's or 1 oz of syrup of ipecac.
2. When vomiting begins, place the victim face down with his or her head lower than the hips. This prevents vomit from entering the airways and causing further damage.

INHALED POISONS

1. Assist or carry the victim to fresh air immediately.
2. Apply artificial respiration if breathing has stopped or is irregular.
3. Call a physician.
4. Treat for shock.
5. Keep the victim as quiet as possible.
6. *Do not* give alcohol in any form to the victim.
7. *Do not* become a victim yourself by exposure to the same poison.
8. The rescuer should employ appropriate protective clothing and breathing apparatus until clear of the hazard.

CHEMICAL CONTAMINATION

Eye Contamination

First aid for chemicals in the eyes involves the immediate washing of the eyes with large quantities of water. Hold the eyelids open and roll the eye while irrigating with water. Emphasis should be placed on the amount of water, the speed with which it is applied, and washing the eye "from the inside outward." Eyes should be washed for at least 10 min. *A delay of 30 sec can mean the difference between no injury to the eye and permanent loss of vision.*

Chemical burns to the eyes can be aggravated by contact lenses. Chemicals spilled in the eyes tend to accumulate under contact lenses. In addition, for proper irrigation, contact lenses need to be removed. It is advisable not to wear contact lenses at a spill site.

In cases of alkaline or acid chemicals in the eyes, irrigation with neutralizing agents should not be used as first-aid treatment. Acids in contact with the cornea will react with protein to form an insoluble barrier. This barrier prevents the acid's penetration into the eye. An alkaline solution does not form this barrier and is free to soak deep into the eye. If this happens with an alkaline solution and an acid neutralizing agent is used, the alkaline solution will be trapped under the insoluble barrier formed by the acid-protein reaction. This will prevent the leaching out of the alkaline solution by irrigation.

Most serious chemical injuries to the eyes can be avoided by quickly and properly washing the eyes with large amounts of water.

Chemicals Spilled on the Body over a Large Area

Quickly remove all contaminated clothing while using the safety shower; seconds count and no time should be wasted because of modesty. Immediately flood the affected body area with cold water for at least 15 min; resume if pain returns. Wash off chemicals by using a mild detergent or soap (preferred) and water; do not use neutralizing chemicals, unguents, or salves.

Chemicals on the Skin in a Confined Area

Immediately flush with cold water and wash with a mild detergent or soap (preferred) and water. If there is no visible burn, scrub with warm water and soap, removing any jewelry in the affected area. If a delayed action [the physiological effects of some chemicals (e.g., methyl and ethyl bromides) may be delayed as much as 48 h] is noted, obtain medical attention promptly and explain carefully what chemicals were involved.

BURNS

General

1. Burns can result from heat (thermal burns) or from chemicals (chemical burns).
2. Shock can complicate every type of burn.
3. A person with "burn shock" may die unless he or she receives immediate first aid.
4. In burn shock, the liquid part of the blood is sent by the body into the burned areas. There may not be enough blood volume left to keep the brain, heart, and other organs functioning normally.
5. All burns should be examined by a physician or nurse.

The objectives of first aid for burns are to:

1. Prevent and treat shock.
2. Prevent contamination.
3. Control pain.

Extensive Thermal Burns

1. Place the cleanest available cloth over all burned body areas to exclude air. The covering for burns should be a clean, thick, dry dressing. Clean newspaper can be substituted if no clean cloth is available.
2. Have the victim lie down.
3. Call a physician.
4. Place the victim's head and chest a little lower than the rest of his or her body. Elevate the legs slightly if possible.
5. If the victim is conscious and can swallow, give him or her plenty of nonalcoholic liquids to drink (water, tea, coffee, dilute salt solution).
6. Move the victim to a hospital immediately.

Small Thermal Burns

1. If the *skin is not broken,* immerse the burned area in clean, cold water to relieve pain and reduce inflammation. Do not apply ice directly to the skin.
2. Soak a sterile gauze pad or clean cloth in a baking soda solution: 2 tablespoonfuls baking soda (sodium bicarbonate) to 1 qt of lukewarm water.
3. Place a pad over the burn and bandage it loosely.
4. *Do not* disturb or open blisters.

Chemical Burns

1. Immediately flush with water; speed in washing is most important in reducing the extent of injury.
2. Flush the affected area with plenty of water.
3. Remove all contaminated clothing and shoes.
4. Place the cleanest available material over the burned area.
5. Treat for shock.
6. If the burned area is extensive, have the victim lie down.
7. Keep him or her down until medical aid is available.
8. Place his or her chest and head a little lower than the rest of the body (raise the legs slightly if possible).
9. Maintain an open airway.
10. If the victim is conscious and can swallow, give him or her plenty of nonalcoholic liquids to drink.
11. *Do not apply ointments,* greases, baking soda, or other substances to extensive burns.

SHOCK

Shock usually accompanies severe injury. It can also follow infection, pain, disturbance of circulation from bleeding, stroke, heart attack, heat exhaustion, food or chemical poisoning, extensive burns, etc.

SIGNS OF SHOCK

1. Cold and clammy skin with beads of perspiration on the forehead and palms of hands
2. Pale face, weakness, dilated pupils, and weak, rapid pulse
3. Complaint by the victim of feeling cold, or even shaking chills
4. Frequent nausea or vomiting
5. Shallow breathing

The following are procedures for the treatment of shock.

TO PREVENT OR MINIMIZE SHOCK

1. If possible, correct the cause of shock (e.g., control bleeding).
2. Keep the victim lying down; if there are no contraindications (e.g., a head injury), elevate the patient's legs.
3. Keep the airway open. If the victim is vomiting, turn his or her head to the side so that the neck is arched.
4. Keep the victim warm if the weather is cold or damp.
5. Give fluids only if the victim does not have head or abdominal injuries, probably will not require surgery, and professional help will be more than 1 h in arriving. Give him or her sips of fluids and do not give stimulants. A suggested formula is 1 pinch of baking soda and 2 pinches of salt per glass (10 oz) of water.
6. Reassure the victim.
7. *Never* give alcoholic beverages to the victim.
8. *Do not* give fluids to unconscious or semiconscious persons.
9. *The prevention of shock should be considered with every injury.*

ENVIRONMENTAL TEMPERATURE EXTREMES

Heat Exhaustion

SYMPTOMS

1. Pale and clammy skin.
2. Rapid and weak pulse.
3. Victim complains of weakness, headache, or nausea.
4. Victim may have cramps in abdomen or limbs.

FIRST AID

1. Have the victim lie down with his or her head level with or lower than his or her body.
2. Move the victim to a cool place, but protect his or her body from chilling.
3. Give the victim salt water (1 teaspoonful of salt to 1 qt of water) to drink *if he or she is conscious.*
4. Loosen tight clothing.
5. Call for medical aid.

Heat Stroke

SYMPTOMS

1. Flushed and hot skin.
2. Rapid and strong pulse.
3. Victim often is unconscious.

FIRST AID

1. Call for medical aid.
2. Cool the victim's body by sponging it with cold water or by cold applications.
3. If the victim is fully conscious and can swallow, give him or her salt water (1 teaspoonful of salt to 1 qt of water).
4. *Do not* give alcohol in any form to the victim.

Frostbite

SYMPTOMS

1. Skin color changes to white or grayish-yellow as frostbite develops.
2. Initial pain that quickly subsides.
3. Victim feels cold and numb; he or she usually is not aware of frostbite.

FIRST AID

1. Cover the frostbitten part of the victim's body with a warm hand or woolen material.
2. If fingers or hand are frostbitten, have the victim hold his hand in his armpit, next to his body.
3. Bring the victim inside as soon as possible.
4. Place the frostbitten area in warm water about 42°C in temperature (108°F).
5. Gently wrap the frostbitten area in blankets if warm water is not available or is impractical to use.
6. Let circulation reestablish itself naturally.
7. When the frostbitten part is warmed, encourage the victim to exercise her fingers and toes.

8. Give the victim a warm, sweet, nonalcoholic drink.
9. *Do not rub* with snow or ice. *Do not use hot water,* hot water bottles, or heat lamps over the frostbitten area.

MISCELLANEOUS ILLNESSES AND INJURIES

After requesting medical aid, the following points should be addressed in specific emergencies:

1. *Abdominal pain:* Keep the patient quiet. Do not give him or her anything by mouth.
2. *Back and neck injuries:* Keep the patient absolutely quiet. Do not move the patient or lift his or her head unless absolutely necessary.
3. *Chest pain:* Keep the patient calm and quiet. Place the patient in the most comfortable position (usually half sitting).
4. *Convulsion or epileptic seizure:* Place the patient on the floor or a couch. Do not restrain the patient's movements except to prevent injury. Do not place a blunt object between the teeth, put any liquid in the mouth, or slap the patient or douse him or her with water.
5. *Fainting:* Simple fainting can usually be ended quickly by laying the victim down.
6. *Unexplained unconsciousness:* Look for emergency medical identification around the victim's neck or wrist or in his or her wallet. Keep the victim warm, lying down, and quiet until he or she regains consciousness. Do not move the victim's head if there is bleeding from the nose, mouth, ear, or eyes. Do not give the victim anything by mouth. Keep the victim's airway open to aid breathing. Do not cramp the neck with a pillow.

MOVING THE INJURED

General

Do not move an injured person until an experienced crew arrives, unless there is real danger of that person receiving further injury by remaining at the accident site.

1. Control bleeding if possible, maintain breathing, and immobilize all suspected fracture sites before moving.
2. Treat for shock.
3. *Pull the victim to safety.*
4. Move the victim's head or feet first, not sideways.
5. Be sure *the victim's* head is protected.

Lifting the Victim to Safety

If the victim must be lifted before a check for injuries can be made, every part of the body should be supported. The body should be kept in a straight line and not bent. Once a victim is lifted, the lifter is responsible for the victim's safe return to the ground or floor.

Exercise care in the approach of any "downed" coworker or bystander. Rapid action may be called for, but hasty and careless intervention may lead to additional injury or loss of life, avoidable if a few moments are taken to assess the immediacy and severity of the situation. Once again, the exercise of careful, informed judgment and plain common sense are the most important safeguards of personnel health.

SUGGESTED READINGS

American Red Cross, *Basic First Aid* (4-vol. set), Doubleday and Co., New York, 1979. Useful for young students.

Lefevre, M., ed., *First Aid Manual for Chemical Accidents,* Van Nostrand Reinhold, New York, 1980.

14
NOISE

INTRODUCTION

School laboratories and workrooms have numerous noise-producing pieces of equipment. In many cases, they cause problems of annoyance, and in more severe situations, they can result in a loss of hearing in those who are chronically exposed to them. Generally, school personnel and students are not exposed to high levels of noise for extended periods of time, but exposure should still be minimized.

WHAT IS NOISE?

By definition, noise is undesired sound. Airborne sound is produced by rapid variations in air pressure above and below atmospheric pressure. The amplitude of these variations determines the intensity of the sound. The pressure variations are averaged (root-mean-square average) and expressed in decibels (dB) in accordance with the formula

$$NdB = 20 \log_{10} (P/P_0)$$

where:
NdB = number of decibels
P = the average (root-mean-square) amplitude of the pressure changes
P_0 = a reference pressure, usually 0.0002 dynes/cm^2

Since the decibel scale is logarithmic, decibels cannot be added directly. The combined level of 90 dB and 90 dB is not 180 dB, but approximately 93 dB. If the values for two sounds are measured separately, the approximate sound pressure level that would result from combining them can be obtained from Table 14-1.

The decibel scale is such that a 10 dB increase represents a 10-fold increase in sound energy. A 20 dB change represents a 100-fold increase, a 30 dB, a 1000-fold increase, and 40 dB, a 10,000-fold energy increase.

Some average sound pressure level readings in dB (referenced to 0.0002 dynes/cm$_2$) are:

Threshold of hearing	0 dB
Conversational speech	60 dB
Office tabulating machine	80 dB
Power wood saw	100 dB
Power lawn mower	105 dB
Wood planer	110 dB
Jet plane	120 dB

The annoyance or hearing loss potential of a noise is not determined only by the total sound pressure level, but also by the distribution of the sound energy with frequency. In obtaining measurements of noise for most purposes, the frequency spectrum is divided into eight bands, each one octave in width. An octave is a frequency interval in which the upper frequency is two times the lower frequency. The octave band frequency limits normally used are 37.5–75, 75–150, 150–300, 300–600, 600–1200, 1200–2400, 2400–4800, and 4800–9600 cycles per second.

WHAT ARE THE EFFECTS OF NOISE?

Excess noise can cause hearing loss and other adverse physiological changes, speech interference, loss of work capacity, and annoyance.

Hearing Damage

Exposure to intense noise will produce hearing loss. This loss may be temporary, permanent, or a combination of the two. Temporary loss or, as it is sometimes called, auditory fatigue will occur after a short exposure to intense noise. If this exposure is continued for a long period of time, it will result in a perma-

TABLE 14-1. Addition of Sound-Level Measurements

Difference Between Levels in dB	Number of dB to Be Added to Higher Level
0	3
2	2
4	1.5
6	1.0
10	0.5

Source: From NIOSH, *Safety in the Laboratory,* 1981, p. IV-51.

nent hearing loss. Some investigators currently working in this area of research believe exposure for 10–12 years will result in a permanent hearing loss equivalent to the temporary loss of hearing experienced after a few hours of exposure to the same noise.

The factors that appear to affect the degree of temporary or permanent loss of hearing are the following:

1. Overall sound pressure level
2. Frequency distribution of this sound energy
3. Total duration of the exposure
4. Time distribution of the exposure
5. Susceptibility of the individual's ears to noise-induced hearing loss

It is believed that any exposure to noise with over a 130-dB sound pressure level is hazardous and should be avoided. At the other extreme, it is believed that no hazard exists for exposure to levels 80 dB and below. For the levels between these two extremes, more information about the exposure is needed, such as the frequency distribution of the sound energy, duration of exposure, and time distribution of exposure.

Most of the noise to which we are exposed is composed of many different frequencies within the audible frequency range, which is approximately 20–20,000 cycles per second. It is known that those sounds containing a large proportion of their energies in the higher-frequency octave bands are more hazardous than those with energy in the low bands. It is also known that if sound energy is concentrated in narrow bands, it is more hazardous than if spread out over wide frequency bands.

The total duration of exposure determines the degree of permanent loss of hearing. A short exposure to noise may cause temporary hearing loss, but no permanent loss. For short exposures to noise, or in the initial exposure stages of prolonged exposures, only temporary hearing loss will develop. With repeated exposures or prolonged exposure time, the ability of the ear to recover totally from temporary hearing loss diminishes. The nonrecoverable or residual loss is the permanent impairment.

Some of the latest research indicates that if the ear is given rest periods from noise exposure, the degree of damage will be reduced. So it appears that an individual who is exposed to noise for, say, 10 min, then is given a 10-min rest before another exposure, can tolerate more noise energy per day than an individual who is exposed continuously for an entire workday. Apparently, the hazard can be reduced by rest periods or the rotation of personnel in and out of intense noise.

There is also a difference in individual susceptibility to noise. Some individuals are able to tolerate much higher exposures to noise than others without showing permanent damage. Much work has been done on tests to determine individual susceptibility in order to select those people who could tolerate high noise exposure. Temporary threshold shift (or sensitivity to noise) after a short noise exposure is being investigated at the present time. This test has not been fully accepted, though, because of insufficient data on the relationship between temporary and permanent hearing loss. Extensive investigations are underway in the field and laboratories to provide a basis for establishing valid and acceptable noise-tolerance criteria for preventing hearing damage. Many damage-risk criteria exist at the present time. It appears that if the sound pressure level in each of the octave bands is maintained below the levels shown in Table 14-2, there is negligible risk of hearing damage even though the exposures may be continuous, 5–8 h a day, for a working lifetime.

If the noise energy is concentrated in narrow bands, limits should probably be 5 dB lower than those presented in Table 14-2. Agreement has not been reached on limits for intermittency noise exposure, and limited information is available on impulse-type noise.

TABLE 14-2. Maximum Allowed Sound Level at Various Frequencies

Octave Band (cps)	Sound Pressure Level (dB)
37.5–75	105
75–150	95
150–300	90
300–600	85
600–1200	85
1200–2400	85
2400–4800	85
4800–9600	85

Source: From NIOSH, *Safety in the Laboratory,* 1981, p. IV-53.

Effects of Noise on Other Physiological Responses

Aside from damage to the hearing mechanism, most noise conditions are not considered to produce other physiological impairments. Sounds of high intensity (over 130 dB) will produce pain in the ear. Exposure to noise above 140 dB may cause dizziness or loss of equilibrium because the balancing organs (semicircular canals) are being stimulated, blurred vision due to the eyeballs fluttering, and may even produce traumatic damage to unprotected ears. There are few situations in which such noise exposure exists. Some examples include operations such as jet engine test cells or missile launch areas.

Speech Interference. The most noticeable effect of noise on man is its effect on speech. A noise not intense enough to produce hearing loss will disrupt communications. Obviously, such disruptions will affect performance in those jobs that depend on reliable verbal communication. This is particularly true on a campus where the major proportion of work is dependent on voice communication, and an unwarranted noise can completely disrupt a classroom or teaching laboratory. Safety problems can also develop due to one's inability to hear commands or danger signals.

Averaging the readings in decibels for the three octave bands 600–1200, 1200–2400, and 2400–4800 cps contained in a broadband noise has been empirically shown to correlate well with the ability to hear speech. The average of these three octave-band dB values is called the speech interference level (SIL). A SIL of 30 dB would be about the maximum that could be tolerated in a medium-size classroom. Small conference rooms are satisfactory with SIL's up to 45–50 dB. At a SIL of 60 dB, telephone use becomes difficult and at 65 dB it becomes difficult to communicate with a raised voice over a distance of 2 ft.

Impairment in Performance. The degrading effects of noise on performance and efficiency would be expected on an intuitive basis. However, there appears to be little evidence that will support this idea. When the studies that have been made on noise and efficiency are examined carefully, it will be found that other factors have entered the study which have affected the results. One factor has been motivation of employees because they believe their employer is providing something for the employees' benefit.

One exception to this is a type of laboratory situation that has shown significant and sustained performance losses under excessive noise where the individual must keep a constant watch over a number of dials or indicators, any one of which may show a change at any given time. This finding has implications for those jobs in automated areas where a worker must continuously monitor or scan a control panel containing many displays that may be indicating an on-going process or a machine malfunction. This decline in alertness could have practical importance for other jobs requiring inspections or the close monitoring of a procedure. In such situations, the number of items being observed at one time must be reduced.

Annoyance. One of the most widespread reactions to noise is annoyance. This is a very difficult problem to evaluate because it is dependent on many factors other than the acoustical stimulus. Some of the nonacoustical factors are:

1. Necessity of the noise
2. Advantage of the noise
3. Inappropriateness of the noise to the activity at hand
4. Unpleasantness associated with the noise
5. Individual tolerance to noise

Some of the acoustical factors are:

1. Loudness
2. Pitch
3. Intermittence and irregularity
4. Localization
5. Time of day

HOW IS NOISE MEASURED?

At the present time, there is a wide assortment of equipment available for noise measurements. This includes sound-survey meters, sound-level meters, octave-band analyzers, narrow-band analyzers, tape and graphic-level recorders, and equipment for the calibration of these instruments. For most noise problems encountered on a campus, the sound-level meter and octave-band analyzer will provide ample information.

Sound-Level Meter

The sound-level meter is the basic instrument for noise measurements. The meter gives a root-mean-square sound pressure level expressed in decibels, referenced to 0.0002 dynes/cm^2. Most meters have three frequency-weighted networks (A, B, and C). The purpose of these is to give a number that is an approximate evaluation of the total loudness level. Human response to sound varies with its frequency and intensity. The ear is less sensitive to the low and high frequencies at low sound intensities. At high sound levels, there is little difference in human response at the various frequencies. The three weighting networks provide a means of compensating for these variations in response. The A network is less sensitive to the low frequencies and is intended for use at sound pressure levels below 55 dB. The B network is an intermediate step for the 55–85 dB range. The C network has a flat response and is used for everything above 85 dB. The values obtained by the use of the three weighting networks will give a measurement of loudness, but for many problems additional information is needed and the sound-level meter must be used with a frequency analyzer. An exception to this is annoyance problems in which measurements on the A scale are commonly used, regardless of the noise's loudness.

The current OSHA permissible maximum sound levels (CFR 1910.95) specify using the A network with a slow response, written as dBA-slow. The resulting maximum allowable noise environment is 85 dBA-slow over an 8-h period, or 100 dBA-slow for 1 h. Sound-level meters specified in the OSHA regulations must meet the requirement of American National Standard Specification, ANSI S1.4 (1971) Type S2A, when used for an A-weighted sound level, slow response.

Octave-Band Analyzer

When the sound to be measured is complex, consisting of a number of tones, or having a continuous spectrum, the single value obtained from a sound-level meter reading often is not sufficient for this purpose. It may be necessary to determine the sound pressure distribution according to frequency. The most practical and widely used analyzer is the octave-band analyzer. As indicated by its name, the upper cutoff frequency for each band is twice the lower cutoff frequency.

The analyzer may be equipped with one of two types of frequency filters. In one type, there is a choice of eight pass-bands and selection is by a single switch. The second type consists of a low cutoff filter, which filters out all frequencies below the setting of the dial, and a high cutoff filter, which filters out all frequencies above the setting of this dial. Measurements can be made in one octave or multiples of octaves up to the complete spectrum with the second analyzer. The standard octaves are 37.5–75, 75–150, 150–300, 300–600, 600–1200, 1200–2400, 2400–4800, and 4800–9600 cycles per second. At the present time, there is a tendency to change the above octaves slightly following the recommendations of the American Standards Association and use octaves with center frequencies of 63, 125, 250, 500, 1000, 2000, 4000, and 8000 cycles per second.

With both the sound-level meter and octave-band analyzer, there is need of equipment for calibration. The calibration equipment usually consists of a signal generator with a small speaker that is placed over a microphone. These calibrators provide a means of calibration in both the laboratory and field.

WHERE ARE NOISE PROBLEMS FOUND?

With the increasing use of power equipment, many potentially hazardous exposures to noise may be added to those that have been present for some time. As a rough guideline to potential noise problems, any noise situation that seriously hampers speech communication at close distances may produce hearing loss if an individual is exposed for long periods of time. Any such noise situation should be investigated with a sound-level meter and octave-band analyzer. Particular attention should be given to length of exposure, as many noise operations may be intermittent with a short total time of exposure per day.

Without noise measurements, it is impossible to say that a particular operation or machine will produce a hazardous noise exposure except for a small group of very intense noise producers, such as jet engines and rockets. The noise level will depend on the speed of machine operation, the maintenance of a machine, and the operating condition of that machine. Table 14-3 provides typical octave-band analyses for some specific pieces of equipment.

NOISE CONTROL?

The Subcommittee on Noise of the Committee of Conservation of Hearing, the American Academy of Ophthalmology and Otolaryngology has outlined a program for the conservation of hearing while exposed to noise that should be very effective. Basically, this is a three-part program.

Analysis of Noise Exposure

Noise exposure is evaluated in terms of:

1. Overall levels
2. Frequency distribution of noise energy

TABLE 14-3. **Typical Overall and Octave Sound Pressure Levels in dB Referenced to 0.0002 dynes/cm²**

			Octave Bands (cps)							
Location	**Remarks**	**Overall**	**20–75**	**75–150**	**150–300**	**300–600**	**600–1200**	**1200–2400**	**2400–4800**	**4800–9600**
Circular cutoff wood saw	cutting 1 in. birch	100	84	84	88	90	91	50	87	84
Planer, wood	furniture parts	108	84	88	95	98	99	97	92	85
Jointer, wood	furniture parts	98	85	88	90	90	89	86	81	75
Milling machine	channel cut steel	90	86	83	78	79	84	85	82	83
Turret lathe	3 in. hole steel	91	81	85	78	80	84	84	82	81
Hand grinder	4 in. electric 1 in. steel	92	85	87	87	4	79	76	76	72
Int. comb. engine test	150 hp	102	86	90	98	?	94	92	90	90
Power house	—	116	109	115	91	79	73	68	68	64
Transonic wind tunnel	Mach 0.3	102	89	79	80	86	90	96	96	88
Supersonic wind tunnel	Mach 2.5	105	86	82	89	98	101	96	90	80
Fuel burner test room	1120 ft/sec	114	89	89	95	103	103	104	105	107
Jet test cell control room	7500 rpm	107	94	94	99	99	98	99	96	89
Aerosol generator	spinning disk 50,000 rpm	101	75	60	62	72	100	80	89	86

Source: From NIOSH, *Safety in the Laboratory,* 1981, p. IV-56.

3. Duration and time distribution of exposure during a typical workday
4. Total exposure time during work life

The measurement of each of these factors of noise exposure is important for hearing conservation. Even though two different noises have the same overall level, their compositions may differ considerably. Also, the auditory effects of continuous noise exposure are different from the effects of intermittent exposure to the same noise.

Control of Noise Exposure

Noise exposure may be reduced by:

1. *Environmental control:*
   (a) Reducing the amount of noise produced at the source.
   (b) Reducing the amount of noise transmitted through air and building structures.
   (c) Revising operational procedures.
2. *Personal protection:*
   The most satisfactory method of environmental control of noise exposure is to control the noise at the source. Unfortunately, this is not always possible. When the amount of noise produced by the source cannot be sufficiently reduced, a combination of control methods may be required to conserve hearing.

Measurement of Hearing

A hearing conservation program should include:

1. Preemployment and/or preplacement hearing tests
2. Routine periodic follow-up tests

These tests provide a record of the initial status of an employee's hearing and make it possible to follow any subsequent changes in hearing ability. Test and retest results will show whether or not the hearing conservation program is effective.

APPENDIX A: HAZARD RATINGS AND CLASSIFICATIONS

TOXICITY RATING SCALE

Rating	Description	Rat Oral, LD_{50} (mg/kg)	Rat 4-h, LC_{50} (ppm)[a]	Probable Lethal Dose for Adult
1	Extremely toxic	<1	<10	Taste (1 grain)
2	Highly toxic	1–50	10–100	4 cc (teaspoon)
3	Moderately toxic	50–500	100–1000	30 g (1 oz)
4	Slightly toxic	500–5000	1000–10,000	250 g (1 pt)
5	Practically nontoxic	5000–15,000	10,000–100,000	1 kg (1 qt)
6	Relatively harmless	>15,000	>100,000	>1 kg (1 qt)

[a]Concentration in air that kills 50% of test animals in 4 h.
Source: Reprinted from Hodge, H. C. and Sterner, J. H., "Tabulation of Toxicity Classes," *American Industrial Hygiene Association Quarterly 10*, 4 (Dec. 1949): 93 by permission of the American Industrial Hygiene Association.

FIRE HAZARD RATINGS AND CLASSES

Class	Criteria	Fire Hazard
IA	Boiling point <100°F Flashpoint >73°F	4
IB	Boiling point at or above 100°F	3
IC	Boiling point not considered Flashpoint 73–100°F	3
II	Flashpoint 100–140°F	2
IIIA	Flashpoint 140–200°F	2
IIIB	Flashpoint >200°F No flashpoint or nonflammable	1 0

UNDERWRITERS LABORATORIES CLASSIFICATIONS

The Underwriters Laboratories classifications grade the relative hazard of various flammable liquids. They are based on the following scale.

Ether class	100
Gasoline class	90–100
Alcohol (ethyl) class	60–70
Kerosene class	30–40
Paraffin oil class	10–20

A flammable liquid is any liquid having a flashpoint below 100°F and a vapor pressure not exceeding 40 lb/in.^{2} (absolute) at 100°F. A combustible liquid is any liquid having a flashpoint at or above 100°F.

APPENDIX B: NATIONAL FIRE PROTECTION ASSOCIATION (NFPA) LABELS

	Identification of Health Hazard, Color Code: Blue		Identification of Flammability, Color Code: Red		Identification of Reactivity (Stability), Color Code: Yellow
	Type of Possible Injury		**Susceptibility of Materials to Burning**		**Susceptibility to Release of Energy**
4	Materials that on very short exposure could cause death or major residual injury even though prompt medical treatment was given.	4	Materials that will rapidly or completely vaporize at atmospheric pressure and normal ambient temperature, or that are readily dispersed in air and that will burn readily.	4	Materials that in themselves are readily capable of detonation or of explosive decomposition or reaction at normal temperatures and pressures.
3	Materials that on short exposure could cause serious temporary or residual injury even though prompt medical treatment was given.	3	Liquids and solids that can be ignited under almost all ambient temperature conditions.	3	Materials that in themselves are capable of detonation or explosive reaction but require a strong initiating source or that must be heated under confinement before initiation or that react explosively with water.
2	Materials that on intense or continuous exposure could cause temporary incapacitation or possible residual injury unless prompt medical treatment is given.	2	Materials that must be moderately heated or exposed to relatively high ambient temperatures before ignition can occur.	2	Materials that in themselves are normally unstable and readily undergo violent chemical change but do not detonate. Also, materials that may react violently with water or that may form potentially explosive mixtures with water.
1	Materials that on exposure would cause irritation but only minor residual injury even if no treatment is given.	1	Materials that must be preheated before ignition can occur.	1	Materials that in themselves are normally stable, but that can become unstable at elevated temperatures and pressures or that may react with water with some release of energy but not violently.

	Identification of Health Hazard, Color Code: Blue		Identification of Flammability, Color Code: Red		Identification of Reactivity (Stability), Color Code: Yellow
	Type of Possible Injury		**Susceptibility of Materials to Burning**		**Susceptibility to Release of Energy**
0	Materials that on exposure under fire conditions would offer no hazard beyond that of ordinary combustible material.	0	Materials that will not burn.	0	Materials that in themselves are normally stable, even under fire exposure conditions, and that are not reactive with water.

FIRE (red)
HEALTH (blue)
SAFETY (yellow)

Source: Reprinted with permission from NFPA 704-1990, *Identification of the Fire Hazards of Materials,* copyright 1990, National Fire Protection Association, Quincy, Mass. This reprinted material is not the complete and official position of the National Fire Protection Association, on the referenced subject which is represented only by the standard in its entirety.

DIMENSIONS OF LABEL

Distances at Which Signals Must be Legible (ft)	Size of Signals Required (in.)
50	1
75	2
100	3
200	4
300	6

Source: Reprinted with permission from NFPA 704-1990, *Identification of the Fire Hazards of Materials,* copyright 1990, National Fire Protection Association, Quincy, Mass. This reprinted material is not the complete and official position of the National Fire Protection Association, on the referenced subject which is represented only by the standard in its entirety.

HAZARD RATINGS AND SIGNAL WORDS

Rating	Health Hazards	Flammability	Reactivity
4	Extreme health hazard	extremely flammable	extremely reactive
3	High health hazard	highly flammable	highly reactive
2	Moderate health hazard	moderately flammable	moderately reactive
1	Slight health hazard	slightly flammable	slightly reactive
0	No significant health hazard	noncombustible	nonreactive

Source: Reprinted with permission from NFPA 704-1990, *Identification of the Fire Hazards of Materials,* copyright 1990, National Fire Protection Association, Quincy, Mass. This reprinted material is not the complete and official position of the National Fire Protection Association, on the referenced subject which is represented only by the standard in its entirety.

APPENDIX C: HAZARDOUS MATERIALS WARNING LABELS

The following information is provided to assist those involved in the storage, handling, and transportation of hazardous materials by presenting potential hazards and precautions associated with the label. The existence of the warning label on packages affords the same quick and easy recognition of hazards to warehouse personnel as is offered to the shippers. For complete details, refer to one or more of the following:

Code of Federal Regulations (CFR), Title 49, Transportation, Parts 100–199. (All modes of transport.) Section numbers listed in this appendix refer to this CFR.

International Civil Aviation Organization (ICAO) Technical Instructions for the Safe Transport of Dangerous Goods by Air. (Air transport.)

International Maritime Organization (IMO) Dangerous Goods Code. (Water transport.)

Canadian Transport Commission (CTC) Regulations. (Rail transport.)

GENERAL GUIDELINES ON THE USE OF LABELS

Labels described in this appendix are used on domestic shipments. Domestic warning labels may display UN class number, division number (and compatibility group for explosives only). [Section 172.407(g) (49 CFR).] Any person who offers a hazardous material for transportation *must* label the package, if required. [Section 172.400(A) (49 CFR).] Label(s), when required, must be printed on or affixed to the surface of the package near the proper shipping name. [Section 172.406(a) (49 CFR).] When two or more different labels are required, display them next to each other. [Section 172.406(c) (49 CFR).] Labels may be affixed to packages (even when not required by regulations), provided each label represents a hazard of the material in the package. [Section 172.401 (49 CFR).] The Hazardous Materials Tables, Section 172.101 and 172.102 (49 CFR), identify the proper label(s) for the hazardous materials listed.

UN CLASS NUMBERS

The following are the hazardous materials class numbers associated with hazard classes:

Class 1 Explosives

Class 2 Gases (compressed, liquefied, or dissolved under pressure)

Class 3 Flammable liquids

Class 4 Flammable solids or substances

Class 5 Oxidizing substances

Class 6 Poisonous and infectious substances

Class 7 Radioactive substances

Class 8 Corrosives

Class 9 Miscellaneous dangerous substances

UN class numbers appear at the bottom of most diamond labels (domestic and international) and placards. See Fig. C-1.

Explosives

UN Class 1 — Examples: A Dynamite
B Propellants or flares
C Common fireworks

Explosive A. Items capable of exploding with a small spark, shock, or flame and spreading the explosion hazard to other packages.

FIGURE C-1. Location of the UN class numbers. (From Department of Defense, *Hazardous Materials, Storage and Handling Handbook,* U.S. Government Printing Office, Washington, D.C., July 1987, p. 82.)

Explosive B. Items are very rapidly combustible.

Explosive C. Items are a low hazard, but may explode under high heat when many are tightly packed together.

Hazards/Precautions. No flares, smoking, flames, or sparks in the hazard area. May explode if dropped, heated, or sparked. See Fig. C-2.

Compressed Gases

UN Class 2 Examples: Acetylene
Chlorine

Items requiring storage and handling under pressure in compressed gas cylinders.

Hazards/Precautions. Container may explode in heat or fire. Contact with liquid may cause frostbite. May be flammable, poisonous, explosive, irritating, corrosive, or suffocating. May be *extremely hazardous.* See Fig. C-3.

FIGURE C-2. Label for explosives, UN class 1. (From Department of Defense, *Hazardous Materials, Storage and Handling Handbook,* U.S. Government Printing Office, Washington, D.C., July 1987, p. 83.)

FIGURE C-3. Label for compressed gases, UN class 2. (From Department of Defense, *Hazardous Materials, Storage and Handling Handbook,* U.S. Government Printing Office, Washington, D.C., July 1987, p. 84.)

Flammable Liquid

UN Class 3 Examples: Ether
Acetone
Gasoline
Toluene
Pentane

Liquids with a flashpoint less than 100°F.

Hazards/Precautions. No flares, smoking, flames, or sparks in the hazard area. Vapors are an explosion hazard. Can be poisonous; check labels. If it is poisonous, it can cause death when inhaled, swallowed, or touched. See Fig. C-4.

Flammable Solid

UN Class 4 Examples: Calcium resinate
Potassium metal
Sodium amide

Any solid material that, under certain conditions, might cause fires or that can be ignited readily and burns vigorously.

FIGURE C-4. Label for flammable liquids, UN class 3. (From Department of Defense, *Hazardous Materials, Storage and Handling Handbook,* U.S. Government Printing Office, Washington, D.C., July 1987, p. 85.)

FIGURE C-5. Label for flammable solids, UN class 4. (From Department of Defense, *Hazardous Materials, Storage and Handling Handbook,* U.S. Government Printing Office, Washington, D.C., July 1987, p. 86.)

Hazards/Precautions. May ignite when exposed to air or moisture. May reignite after extinguishing. Fires may produce irritating or poisonous gases. Contact may cause burns to skin or eyes. See Fig. C-5.

Dangerous When Wet

UN Class 4	Examples:	Magnesium metal
		Aluminum phosphide
		Lithium hydride
		Calcium carbide

These items include flammable solids that are reactive with water.

Hazards/Precautions. May ignite in the presence of moisture. May reignite after fire is extinguished. Contact may cause burns to skin and eyes. Skin contact may be poisonous. Inhalation of vapors may be harmful. Prohibit flames or smoking in area. See Fig. C-6.

Oxidizing Material

UN Class 5	Examples:	Calcium permanganate
		Calcium hypochlorite
		Barium perchlorate
		Hydrogen peroxide
		Ammonium nitrate

FIGURE C-6. Label for materials that are dangerous when wet, UN class 4. (From Department of Defense, *Hazardous Materials, Storage and Handling Handbook,* U.S. Government Printing Office, Washington, D.C., July 1987, p. 87.)

These items are chemically reactive and will provide both heat and oxygen to support a fire.

Hazards/Precautions. May ignite combustibles (wood, paper, etc.). Reaction with fuels may be violent. Fires may produce poisonous fumes. Vapors and dusts may be irritating. Contact may burn skin and eyes. Peroxides may explode from heat or contamination. See Fig. C-7.

Poisonous Material

Class B poison	Examples:	Cyanogen gas
UN Class 6		Lead cyanide
		Parathion

Items are extremely toxic to man and animals.

Hazards/Precautions. May cause death quickly if breathed, swallowed, or touched. May be flammable, explosive, corrosive, or irritating. May be *extremely hazardous.* Look for the "skull and crossbones" on the label. The degree of hazard key words:

Poison	Highly toxic
Danger	Moderately toxic
Warning	Least toxic

Read the label carefully for storage and safety information. See Fig. C-8.

Irritating Material

UN Class 6	Examples:	Tear gas
		Riot control agent

Items capable of causing discomfort such as tearing, choking, vomiting, and skin irritation.

Hazards/Precautions. May cause difficulty in breathing. May burn but do not ignite readily. Exposure in enclosed areas may be harmful. May cause

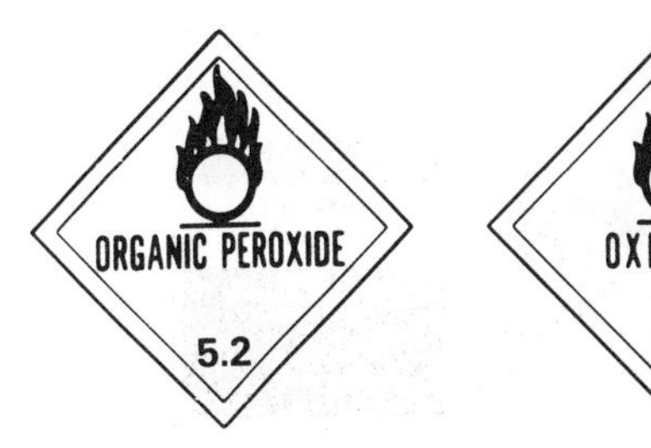

FIGURE C-7. Label for oxidizing materials, UN class 5. (From Department of Defense, *Hazardous Materials, Storage and Handling Handbook,* U.S. Government Printing Office, Washington, D.C., July 1987, p. 88.)

FIGURE C-8. Label for poisonous materials, Class B poisons, UN class 6. (From Department of Defense, *Hazardous Materials, Storage and Handling Handbook,* U.S. Government Printing Office, Washington, D.C., July 1987, p. 89.)

tearing of the eyes, choking, nausea, or skin irritation. See Fig. C-9.

Biomedical Material

UN Class 6 Examples: Live virus vaccines
Etiologic agents, n.o.s.

Items that can cause human disease (infectious/etiological agent).

Hazards/Precautions. May be ignited if the carrier is flammable. Contact may cause infection or disease. Damage to outer container may not affect inner container. See Fig. C-10.

FIGURE C-9. Label for irritating materials, UN class 6. (From Department of Defense, *Hazardous Materials, Storage and Handling Handbook,* U.S. Government Printing Office, Washington, D.C., July 1987, p. 90.)

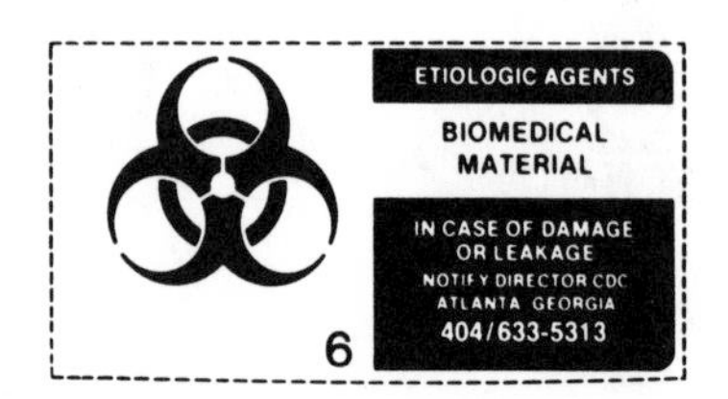

FIGURE C-10. Label for biological hazards, UN class 6. (From Department of Defense, *Hazardous Materials, Storage and Handling Handbook,* U.S. Government Printing Office, Washington, D.C., July 1987, p. 91.)

Radioactive Material

UN Class 7 Examples: Thorium 232
Carbon 14
Radium 226

Degree of hazard will vary depending on type and quantity of material.

Hazards/Precautions. Avoid touching broken or damaged radioactive items.

PERSONS HANDLING DAMAGED ITEMS MUST WEAR RUBBER OR PLASTIC GLOVES.

Damaged items will be monitored and safely packaged under the surveillance of the radiological monitor. Persons having come in direct contact with damaged or broken radioactive items will move away from the spill site (but stay in the area) to be monitored and decontaminated. See Fig. C-11.

Corrosives

UN Class 8 Examples: Sodium hydroxide
Hydrochloric acid
Alkaline liquids

Items include materials that cause destruction to human tissue and corrode metal (i.e., steel) and thus many packaging materials upon contact.

Hazards/Precautions. Contact causes burns to skin and eyes. May be harmful if breathed. Fire may produce poisonous fumes. May react violently with water. May ignite combustibles. Explosive gases may accumulate. See Fig. C-12.

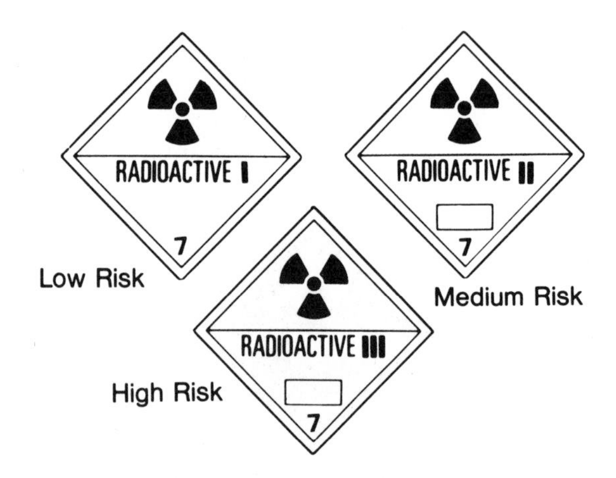

FIGURE C-11. Label for radioactive materials, UN class 7. (From Department of Defense, *Hazardous Materials, Storage and Handling Handbook,* U.S. Government Printing Office, Washington, D.C., July 1987, p. 92.)

FIGURE C-12. Label for corrosives, UN class 8. (From Department of Defense, *Hazardous Materials, Storage and Handling Handbook,* U.S. Government Printing Office, Washington, D.C., July 1987, p. 93.)

Other Regulated Material (ORM) Markings

(49 CFR, Part 172.316) Applies to material that may pose an unreasonable risk to health and safety or property, and is not covered under the hazardous materials warning label requirements.

ORM-A. Material with an anesthetic, irritating, noxious, toxic, or other properties that can cause discomfort to persons in the event of leakage. See Fig. C-13. Examples include trichloroethylene, 1,1,1-trichloroethane, dry ice, chloroform, carbon tetrachloride.

ORM-B. Material specifically named or capable of causing significant corrosion damage from leakage. See Fig. C-14. Examples include lead chloride, quicklime, metallic mercury, barium oxide.

ORM-C. Material specifically named and with characteristics that make it unsuitable for shipment unless properly packaged. See Fig. C-15. Examples include bleaching powder, lithium batteries (for disposal), magnetized materials, sawdust, asbestos.

FIGURE C-13. ORM-A label. (From Department of Defense, *Hazardous Materials, Storage and Handling Handbook,* U.S. Government Printing Office, Washington, D.C., July 1987, p. 94.)

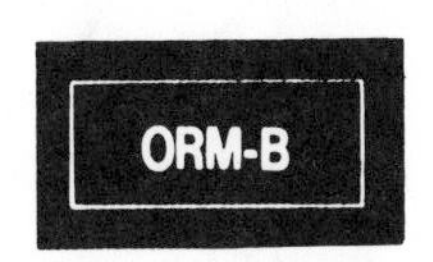

FIGURE C-14. ORM-B label. (From Department of Defense, *Hazardous Materials, Storage and Handling Handbook,* U.S. Government Printing Office, Washington, D.C., July 1987, p. 95.)

FIGURE C-15. ORM-C label. (From Department of Defense, *Hazardous Materials, Storage and Handling Handbook,* U.S. Government Printing Office, Washington, D.C., July 1987, p. 95.)

ORM-D. Material such as a consumer commodity that presents a limited hazard due to form, quantity, and packaging. A material for which an exception must be provided. See Fig. C-16. Examples include chemical consumer commodities (e.g., hairspray and shaving lotion) and small arm ammunition (reclassified because of packaging).

ORM-E. Material that is not included in any other hazard class, but is regulated as ORM. See Fig. C-17. Examples include hazardous waste and hazardous substances.

Special Handling

Additional labeling to be used with DOT hazardous materials warning labels, as required (49 CFR, Part 172.402). See Fig. C-18. An example is asbestos (29 CFR, Part 1910.1001).

CAUTION
Contains Asbestos Fibers.
Avoid Creating Dust.
Breathing Asbestos Dust May Cause
Serious Bodily Harm.

FIGURE C-16. ORM-D label. (From Department of Defense, *Hazardous Materials, Storage and Handling Handbook,* U.S. Government Printing Office, Washington, D.C., July 1987, p. 96.)

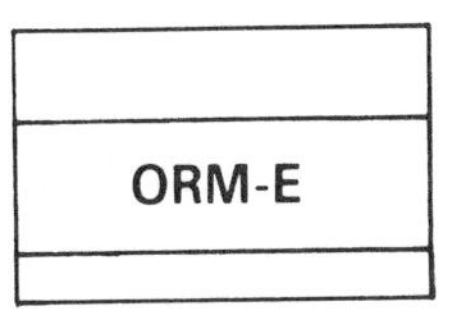

FIGURE C-17. ORM-E label. (From Department of Defense, *Hazardous Materials, Storage and Handling Handbook,* U.S. Government Printing Office, Washington, D.C., July 1987, p. 96.)

Cargo Aircraft only Label

Magnetized Material Label

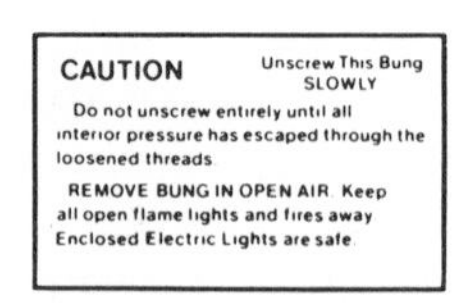

Bung Label

Empty Label

FIGURE C-18. Examples of labels describing special handling. (From Department of Defense, *Hazardous Materials, Storage and Handling Handbook,* U.S. Government Printing Office, Washington, D.C., July 1987, p. 97.)

FIGURE C-19. Special-item labels. (From Department of Defense, *Hazardous Materials, Storage and Handling Handbook,* U.S. Government Printing Office, Washington, D.C., July 1987, p. 98.)

Special-Item Markings

Examples include PCB's (40 CFR, Part 761.40) and hazardous waste (40 CFR, Part 262.32). See Fig. C-19.

MANUFACTURER'S LABELS

Hazard Communication Label (29 DFR, Part 1910.1200)

Applies to the hazards of all chemicals manufactured or imported.

Ensures that information on chemical hazards is properly transmitted.

Label contents: Product identity, hazard warnings, and name and address of the manufacturer, importer, or other responsible party.

Pesticides (40 CFR, Part 162.10)

Applies to all pesticide products.

Label contents: Product identity, producer/registrant data, net contents, product registration number, ingredients statement, warning or precautionary statement, directions for use, and use classification.

Substance liable to Spontaneous Combustion

Poisonous Substance

Poisonous Substance

Infectious Substance

FIGURE C-20. Examples of international labels. (From Department of Defense, *Hazardous Materials, Storage and Handling Handbook,* U.S. Government Printing Office, Washington, D.C., July 1987, p. 100.)

INTERNATIONAL LABELING

Examples of international labels are shown in Fig. C-20. These are examples of international labels not presently used for domestic shipments. Most domestic labels may be used internationally. The text of such labels when used internationally may be in the language of the country of origin. Text is mandatory on radioactive material, St. Andrew's Cross, and infectious substance labels.

International Labeling for Explosives

Examples of such international explosive labels are given in Fig. C-21. The *numerical designation* represents the *class* or *division*. The *alphabetical designation* represents the *compatibility group* (for explosives only). *Division numbers* and *compatibility group* combinations can result in over 30 different "explosives" labels (see Technical Instructions for the Safe Transport of Dangerous Goods by Air, the International Civil Aviation Organization.)

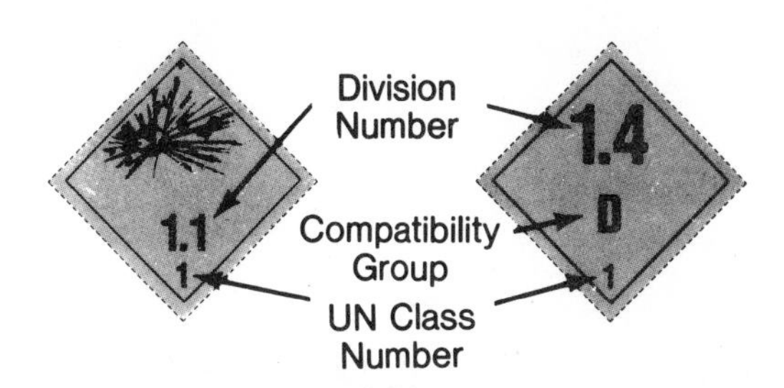

FIGURE C-21. International explosives labels. (From Department of Defense, *Hazardous Materials, Storage and Handling Handbook,* U.S. Government Printing Office, Washington, D.C., July 1987, p. 101.)

Suggested Reading

Hazardous Materials Storage and Handling Handbook, U.S. Government Printing Office, Washington, D.C., 1987.

APPENDIX D: HAZARDOUS LABORATORY SUBSTANCES

Assessment of the chemicals in the following table indicates that their hazardous nature is greater than their potential usefulness in many secondary school science programs. Evaluation included toxicity, carcinogenicity, teratogenicity, flammability, and explosive propensity. These chemicals should be removed from schools if alternatives can be used. For those that must be retained, amounts should be kept to a minimum. Irritants and corrosives can be used with caution and approved safety procedures.

Chemical Name	CAS#	Hazardous Properties
Acetaldehyde	75-07-0	IARC 3; irritant
Acetamide		remove if feasible; animal carcinogen
Acetic acid		corrosive
Acetic anhydride		irritant
N-Acetoxy-2-acetylaminofluorene (N-AcO-AAF)		potential occupational carcinogenic risk
2-Acetylaminofluorene	53-96-3	NTPAHC
Acid green		remove if feasible
Acrylamide	79-06-1	IARC 3
Acrylonitrile	107-13-1	IARC 2A; NTPAHC; too hazardous
Actinolite	13768-00-8	IARC 1; NTPHC
Actinomycin D	50-76-0	IARC 2B
Aflatoxins	1402-68-2	IARC 2A; NTPAHC
Aflatoxin B1	1162-65-8	IARC 2A
Aflatoxin B_2		potential occupational carcinogenic risk
Aflatoxin G_1		potential occupational carcinogenic risk
Aflatoxin G_2		potential occupational carcinogenic risk
Aflatoxin M_1		potential occupational carcinogenic risk
Aflatoxin M_2		potential occupational carcinogenic risk
Aflatoxin P_1		potential occupational carcinogenic risk
AF-2	3688-53-7	IARC 3
Aldrin		suspected carcinogen
Aliphatic halides		potential occupational carcinogenic risk
Aluminum chloride		remove if feasible; fire hazard; corrosive
2-Aminoanthraquinone	117-79-3	NTPAHC
o-Aminoazotoluene	97-56-3	IARC 2Bs; PAFAR
4-Aminobiphenyl	92-67-1	IARC 1; NTPHC
2-Aminodipyrido[1,2-a:3′,2′-d]dimidazole	67730-10-3	IARC 3
2-Aminofluorene (AF)		potential occupational carcinogenic risk
Aminofluorene derivatives		potential occupational carcinogenic risk
2-Amino-6-methyldipyrido[1,2-a:3′,2′-d]imidazole	67730-11-4	IARC 3
2-Amino-3-methyl-9H-pyrido[2,3-b]indole	68006-83-7	IARC 3
Ammonium bichromate		remove if feasible; toxic; fire hazard; corrosive
Ammonium chromate		too hazardous
Ammonium oxalate		remove if feasible; corrosive
Ammonium vanadate		remove if feasible; toxic
Amosite	12172-73-5	IARC 1; NTPHC
Aniline		too hazardous; toxic
Aniline hydrochloride		too hazardous; toxic
o-Anisidine	90-04-0	IARC2Bs; NTPAHC

Chemical Name	CAS#	Hazardous Properties
o-Anisidine hydrochloride	134-29-2	NTPAHC
Anthophyllite	17068-78-9	IARC 1; NTPHC
Anthracene		too hazardous; carcinogen
Antimony		remove if feasible; toxic; irritant
Antimony chloride		toxic; corrosive; reactive hazard
Antimony oxide		remove if feasible; toxic
Antimony pentachloride		corrosive
Antimony potassium tartrate		remove if feasible; toxic
Antimony trichloride		too hazardous; corrosive
Antimony trioxide		suspected carcinogen
Antracene oils		SEQ NO-2-1 IARC 3
Arsenic compounds	7440-38-2	NTPHC
Arsenic pentoxide	1303-28-2	IARC 1; too hazardous; toxic
Arsenic trioxide	1327-53-3	IARC 1; NTPHC; too hazardous
Arsenic, inorganic compounds	7440-38-2	IARC 1; NTPHC
Aryl halides		potential occupational carcinogenic risk
Asbestos	1332-21-4	IARC 1; NTPHC; too hazardous
Asbestos, amosite		carcinogen
Asbestos, chrysotile		carcinogen
Asbestos, crocidolite	12001-28-4	IARC 1; NTPHC
Ascarite		too hazardous; carcinogen
Auramine	492-80-8	IARC 2B; NTPHC
Auramine, technical grade	492-80-8	IARC 2B
Azaserine	115-02-6	IARC 2Bs
Azathioprine	446-86-6	IARC 1; NTPHC
Aziridines		potential occupational carcinogenic risk
Azo and azoxy derivatives		potential occupational carcinogenic risk
Barium chloride		remove if feasible; toxic
Barium chromate	10294-40-3	IARC 1
Barium compounds		toxic
Barium hydroxide		toxic
Benzo(a)pyrene	50-32-8	IARC 2A; NTPAHC
Benzo(b)fluoranthene	205-99-2	IARC 2Bs; NTPAHC
Benzo(j)fluoranthene	205-82-3	IARC 3; PAFAR
Benzo(k)fluoranthene	207-08-9	IARC 3; PAFAR
Benzone		remove if feasible
Benzotrichloride	98-07-7	IARC 2B; NTPAHC
Benzoyl peroxide		too hazardous; toxic, reactive hazard; explosive
Benzyl violet	1694-09-3	IARC 2Bs
Beryl	1302-52-9	IARC 2A; NTPAHC
Beryllium	7440-41-7	IARC 2A; NTPAHC
Beryllium-aluminum alloy	12770-50-2	NTPAHC; IARC 2A
Beryllium carbonate	66104-24-3	IARC 2A; NTPAHC; remove if feasible; toxic
Beryllium chloride	7787-47-5	IARC 2A; NTPAHC
Beryllium compounds	7440-41-7	IARC 2A; NTPAHC
Beryllium fluoride	7787-49-7	IARC 2A; NTPAHC
Beryllium hydrogen phosphate	13598-15-7	IARC 2A; NTPAHC
Beryllium hydroxide	13327-32-7	IARC 2A; NTPAHC
Beryllium oxide	1304-56-9	IARC 2A; NTPAHC
Beryllium silicate	13598-00-0	IARC 2A; NTPAHC
Beryllium sulfate	13510-49-1	IARC 2A; NTPAHC
Bitumens		SEQ NO-3-9 IARC 3
Bonine fluid		remove if feasible
Bromine		remove if feasible; corrosive; toxic
Bromkal 80	61288-13-9	IARC 3
Bromoethyl methanesul fonate (BEMS)		potential occupational carcinogenic risk
1,3-Butadiene	106-99-0	IARC 3; PAFAR
1,4-Butanediol dimethanesulphonate	55-98-1	IARC 1; NTPHC
Butanol		Toxic
Butylated hydroxyanisole	25013-16-5	IARC 3
β-Butyrolactone	3068-88-0	IARC 2Bs
C.I. basic red 9 (*p*-rosaniline)	569-61-9	PAFAR
C.I. direct black 38	1937-37-7	IARC 2B; NTPAHC

Chemical Name	CAS#	Hazardous Properties
C.I. direct blue 6	2602-46-2	NTPAHC
C.I. direct blue 14	72-57-1	IARC 2Bs
C.I. direct disperse orange 11	82-28-0	NTPAHC
C.I. solvent orange 2	2646-17-5	IARC 2Bs
Cadmium	7440-43-9	NTPAHC; IARC 2B; remove if feasible
Cadmium acetate		remove if feasible; toxic
Cadmium bromide		remove if feasible; toxic
Cadmium carbonate		remove if feasible; toxic
Cadmium sulfate	10124-36-4	IARC 2B; NTPAHC; remove if feasible
Cadmium sulfide	1306-23-6	IARC 2B; NTPAHC
Calcium carbide		corrosive
Calcium chromate, sintered	13765-19-0	IARC 1; NTPHC
Calcium cyanide		too hazardous; toxic
Calcium fluoride		too hazardous; irritant
Carbon blacks, solvent (benzene) extracts	SEQ NO-40-6	IARC 3
Carbon disulfide		explosive
Carbon tetrachloride	56-23-5	IARC 2B; NTPAHC; too hazardous
Carboxylic acid derivatives		potential occupational carcinogenic risk
Carmine		remove if feasible
Carrageenan (degraded)	9000-07-1	IARC 3
Catechol		remove if feasible; toxic; irritant
Chloral hydrate		too hazardous; toxic
Chlorambucil	305-03-3	IARC 1; NTPHC
Chloramphenicol	56-75-7	IARC 2B
Chlorendic acid	115-28-6	PAFAR
Chloretone		too hazardous; toxic
Chlorinated ethanes		suspected carcinogen
Chlorinated toluenes, production of	SEQ NO-39-8	IARC 2B
Chloroform	67-66-3	IARC 2B; NTPAHC; too hazardous
Chloromethyl ethers		potential occupational carcinogenic risk
Chloromethyl methyl ether, technical grade	107-30-2	IARC 1; NTPHC
4-Chloro-2-methyl-phenoxy acetic acid (occupational exposure)	94-74-6	IARC 2Bs
2-(4-Chloro-2-methyl-phenoxy)propanoic acid (occupational exposure)	93-65-2	IARC 2Bs
3-Chloro-2-methylpropene	563-47-3	PAFAR
Chlorophenols (occupational exposure)	SEQ NO-5-8	IARC 2B; IARC 2Bs
4-Chloro-*o*-phenylenediamine	95-83-0	IARC 2Bs; NTPAHC
Chloropromazine		too hazardous; toxic
p-Chloro-*o*-toluidine	95-69-2	IARC 3
Chromates		suspected carcinogen
Chromic acid		remove if feasible; corrosive; reactive hazard
Chromic acid, calcium salt (1:1)	13765-19-0	IARC 1; NTPHC
Chromic acid, calcium salt (1:1), dihydrate	8012-75-7	IARC 1
Chromium	7440-47-3	IARC 1; NTPHC; too hazardous
Chromium acetate		remove if feasible
Chromium compounds	7440-47-3	IARC 1; NTPHC
Chromium(III) oxide		suspected carcinogen
Chromium oxide		too hazardous
Chromium potassium sulfate		too hazardous
Cisplatin	15663-27-1	IARC 2B
Citrus red no. 2	6358-53-8	IARC 2Bs
Coal gasification	SEQ NO-40-3	IARC 2Bs
Coal soot (occupational exposure)	SEQ NO-7-0	IARC 1; NTPHC
Coal soot extracts	SEQ NO-7-0	IARC 3
Coal tar pitches	65996-93-2	IARC 2Bs
Coal tar pitches (occupational exposure)	65996-93-2	IARC 1
Coal tars	8007-45-2	IARC 3
Coal tars (occupational exposure)	807-45-2	IARC 1
Cobalt		suspected carcinogen; remove if feasible
Cobalt nitrate		remove if feasible; toxic; reactive hazard
Cobalt-chromium alloy	11114-92-4	IARC 1; NTPHC
Coke oven emissions	SEQ NO-7-5	NTPHC

Chemical Name	CAS#	Hazardous Properties
Coke production	SEQ NO-40-4	IARC 2Bs
Colchicine		suspected carcinogen; too hazardous; toxic
Creosote(s)	8001-58-9	IARC 2Bs
Creosote(s) (occupational exposure)	8001-58-9	IARC 1; NTPHC
p-Cresidine	120-71-8	IARC 2Bs; NTPAHC
Cummingtonite-grunerite	SEQ NO-37-3	NTPHC
Cupferron	135-20-6	NTPAHC
Cycasin	14901-08-7	IARC 2Bs; NTPAHC
Cyclohexane		remove if feasible; toxic; fire hazard
Cyclohexene		remove if feasible; toxic; fire hazard
Cyclophosphamide	50-18-0	IARC 1; NTPHC
Dacarbazine	4342-03-4	IARC 2B; NTPAHC
Daunomycin	20830-81-3	IARC 2Bs
DDT	50-29-3	IARC 2B; NTPAHC
Decabromobiphenyl	13654-09-6	IARC 3
Dehydrosafrole		suspected carcinogen
Dethylene oxide	75-21-8	NTPAHC
N,N′-diacetylbenzidine	613-35-4	IARC 2Bs
Diallate		suspected carcinogen
2,4-Diaminoanisole sulfate	39156-41-7	IARC 2Bs; NTPAHC
4,4′-Diaminodiphenyl ether	101-80-4	IARC 2Bs; PAFAR
2,4-Diaminotoluene	95-80-7	IARC 2Bs; NTPAHC
Diazomethane		potential occupational carcinogenic risk
Dibenz(a,h)acridine	226-36-8	IARC 2Bs; NTPAHC
Dibenz[a,j]acridine	224-42-0	IARC 2Bs; NTPAHC
Dibenz[a,h]anthracene	53-70-3	IARC 2Bs; NTPAHC
Dibenzo(a,e)pyrene	192-65-4	IARC 2Bs; PAFAR
1,2-Dibromo-3-chloropropane (DBCP)	96-12-8	IARC 2Bs; NTPAHC
1,2-Dibromoethane		suspected carcinogen
p-Dichlorobenzene	106-46-7	PAFAR; irritant
3,3′-Dichlorobenzidine	91-94-1	IARC 2B; NTPAHC
3,3′-Dichloro-4,4′-diaminodiphenyl ether	28434-86-8	IARC 2Bs
1,2-dichloroethane		suspected carcinogen
2,4-Dichlorophenol (occupational exposure)	120-83-2	IARC 2Bs
2,4-Dichlorophenoxy acetic acid (occupational exposure)	94-75-7	IARC 2Bs
2-(2,4-Dichlorophenoxy)-propanoic acid (occupational exposure)	120-36-5	IARC 2Bs
1,3-Dichloropropene	542-75-6	IARC 3
1,2-Diethylhydrazine	1615-80-1	IARC 2Bs
Dichlorobenzene		too hazardous
Dichloroethane		too hazardous
Dichloroindophenol, sodium salt		remove if feasible
Dichloropropene	542-75-6	PAFAR
Dieldrin		suspected carcinogen
Dienestrol	84-17-3	IARC 2B
Diepoxybutane (DEB)	1464-53-5	IARC 2Bs; NTPAHC
Diethyl phthalate		irritant
Diethyl sulfate	64-67-5	IARC 2A; NTPAHC
4-Dimethylaminoazobenzene		potential occupational carcinogenic risk
trans-2((Dimethylamino)methylimino)-5-(2-(5-nitro-2-furyl)vinyl)-1,3,4-oxadiazole	55738-54-0	IARC 2Bs
N,N-dimethylaniline		Toxic
7,12-Dimethylbenz[a]anthracene (DMBA)		potential occupational carcinogenic risk
3,3′-Dimethylbenzidine		potential occupational carcinogenic risk
1,1-Dimethylethylenimine		potential occupational carcinogenic risk
1,1-Dimethylhydrazine (UDMH)	57-14-7	IARC 2Bs; NTPAHC
1,2-Dimethylhydrazine (SDMH)	540-73-8	IARC 2Bs
Dimethyl sulfate	77-78-1	IARC 2A; NTPAHC
Dimethylaminoazobenzene	60-11-7	IARC 2Bs; NTPAHC
Dimethylaniline		too hazardous
Dimethylcarbamoyl chloride	79-44-7	IARC 2B; NTPAHC
Dimethylvinyl chloride	513-37-1	PAFAR

Chemical Name	CAS#	Hazardous Properties
2,4-Dinitrophenol		remove if feasible; toxic; explosion hazard
1,4-Dinitrosopiperzine (DNP)		potential occupational carcinogenic risk
2,4-Dinitrotoluene		suspected carcinogen
Dioxane	123-91-1	IARC 2B; NTPAHC
p-Dioxane		suspected carcinogen; too hazardous
Diphenyl ester carbonic acid		suspected carcinogen; too hazardous
1,2-Diphenylhydrazine	122-66-7	NTPAHC
Estrogens, conjugated	SEQ NO-24-0	IARC 1; NTPHC
Estrone	53-16-7	IARC 2B; NTPAHC
Ethenylamine, N-methyl-N-nitroso	4549-40-0	IARC 2Bs; NTPAHC
Ethers, oxides and epoxides		potential occupational carcinogenic risk
Ethinylestradiol	57-63-6	IARC 2B; NTPAHC
Ethionine		potential occupational carcinogenic risk
Ethyl acrylate	140-88-5	IARC 3; PAFAR
Ethyl carbamate	51-79-6	IARC 2Bs; NTPAHC
Ethyl ether		fire hazard; explosion hazard
Ethyl methacrylate		irritant
Ethyl methanesulfonate (EMS)	62-50-0	IARC 2Bs
Ethylene bis dithiocarbamate		suspected carcinogen
Ethylene dibromide (EDB)	106-93-4	IARC 2B; NTPAHC
Ethylene dichloride	107-06-2	IARC 2Bs; NTPAHC; too hazardous
Ethylenimine (EI)		potential occupational carcinogenic risk
Ethylene oxide	75-21-8	IARC 2B; too hazardous; toxic; fire hazard
Ethylene thiourea	96-45-7	IARC 2B; NTPAHC
Ferric chloride		irritant
Ferrous sulfate		remove if feasible
Firemaster BP-6	59536-65-1	IARC 3; NTPAHC
2-(2-Formylhydrazino)-4-(5-nitro-2-furyl)thiazole	3570-75-0	IARC 2Bs
Fuchsin		remove if feasible
Furniture manufacture	SEQ NO-40-0	IARC 1
Gasoline		remove if feasible; fire hazard
Glycidylaldehyde	765-34-4	IARC 2Bs
Gunpowder		too hazardous; explosion hazard
Gyromitrin	16568-02-8	IARC 3
Hematite underground mining, with exposure to radon	1317-60-8	IARC 1; NTPHC
Hematoxylin		remove if feasible
Heptachlor		suspected carcinogen
2,2′,4,4′,5,5′-Hexabromo-1,1′-biphenyl	59080-40-9	IARC 3
Hexachlorobenzene	118-74-1	IARC 2Bs; NTPAHC
Hexachlorobutadiene		suspected carcinogen
Hexachlorocyclohexane		suspected carcinogen
Hexachlorophene		too hazardous; toxic; irritant
Hexamethyl phosphoramide		suspected carcinogen
Hexamethylphospharamide	680-31-9	NTPAHC; IARC 2Bs
Hydrazine (HZ)	302-01-2	IARC 2B; NTPAHC
Hydrazines		potential occupational carcinogenic risk
Hydrazine, sulfate (1:1)	10034-93-2	NTPAHC
Hydrogen peroxide (30%)		irritant
Hydrogen sulfide		remove if feasible; irritant; fire hazard
Hydroiodic acid		corrosive
Hydroquinone		remove if feasible; irritant
N-Hydroxy-2-acetylaminofluorene (N-HO-AAF)		potential occupational carcinogenic risk
Indeno(1,2,3-cd)pyrene	193-39-5	IARC 2Bs; NTPAHC
Indigo carmine		too hazardous
Iodine crystals		toxic; irritant
2-amino-3-methylimidazo[4,5-f]quinoline	76180-96-6	IARC 3
Iron and steel founding	SEQ NO-40-5	IARC 2Bs
Iron dextran	9004-66-4	IARC 3; NTPAHC
Iso-amyl alcohol		remove if feasible
Iso-butyl alcohol		remove if feasible
Iso-pentyl alcohol		remove if feasible
Isopropyl alcohol manufacture (strong acid process)	67-63-0	IARC 1; NTPHC
Isosafrole	120-58-1	IARC 2Bs

Chemical Name	CAS#	Hazardous Properties
Kepone	143-50-0	IARC 2Bs; NTPAHC
Lasiocarpine	303-34-4	IARC 2Bs
Lead acetate	301-04-2	IARC 3; NTPAHC
Lead arsenate		suspected carcinogen; too hazardous
Lead subacetate	1335-32-6	IARC 3
Lead(VI) chromate		too hazardous
Lindane and other hexachlorocyclohexane isomers	58-89-9	NTPAHC
α-Lindane	319-84-6	NTPAHC
β-Lindane	319-85-7	NTPAHC
γ-Lindane	58-89-9	NTPAHC
Lindane-mixed isomers	608-73-1	NTPAHC
Lithium		toxic; fire hazard; explosion hazard; corrosive; too hazardous
Lithium nitrate		too hazardous
Magenta, manufacture of	632-99-5	IARC 2A
Magnesium		fire hazard
Magnesium, metal (powder)		too hazardous
Magnesium chlorate		remove if feasible; toxic; reactive hazard; fire hazard
Melphalan	148-82-3	IARC 1; NTPHC
Mercuric sulfide		toxic
Mercuric bichloride		remove if feasible
Mercuric chloride		toxic; too hazardous
Mercuric iodide		remove if feasible; toxic
Mercurous nitrate		remove if feasible; toxic
Mercurous oxide		remove if feasible; toxic
Mercury		too hazardous; toxic
Merphalan	531-76-0	IARC 2Bs
Mesitylene		too hazardous; toxic
Mestranol	72-33-3	IARC 2B; NTPAHC
5-methoxypsoralen	484-20-8	IARC 3
3′-Methyl-4-aminoazobenzene		potential occupational carcinogenic risk
3-Methylcholanthrene (MC)		potential occupational carcinogenic risk
Methyl chloromethyl ether	107-30-2	NTPHC
5-Methylcrysene	3697-24-3	IARC 3; PAFAR
4,4′-Methylene bis-(2-chloroaniline)	101-14-4	IARC 2Bs; NTPAHC
4,4′-Methylene bis(N,N-dimethyl)benzeneamine	101-61-1	NTPAHC
4.4′-Methylene bis(2-methylaniline)	838-88-0	IARC 2Bs
4,4′-Methylenedianiline	101-77-9	IARC 3; NTPAHC
4,4′-Methylenedianiline dihydrochloride	13552-44-8	IARC 3; NTPAHC
Methyl ethyl ketone		remove if feasible; toxic; irritant; fire hazard; explosion hazard
Methyl hydrazine (NMH)		potential occupational carcinogenic risk
Methyl iodide	74-88-4	IARC 2Bs; NTPAHC
2-Methyl-1-nitroanthraquinone	129-15-7	IARC 2Bs
N-Methyl-N′-nitro-N-nitrosoguanidine	70-25-7	IARC 2Bs
Methyl oleate		remove if feasible
Methyl orange[a]		too hazardous; toxic
Methyl red[a]		too hazardous
Methyl salicylate		irritant
Methylazoxymethanol acetate	592-62-1	IARC 2Bs
Methylene chloride	75-09-2	IARC 3; PAFAR
Methylthiouracil	56-04-2	IARC 2Bs
Metronidazole	443-48-1	IARC 2B; NTPAHC
Michler's ketone	90-94-8	NTPAHC
Mirex	2385-85-5	IARC 2Bs; NTPAHC
Mitomycin C	50-07-7	IARC 2Bs
Monocrotaline	315-22-0	IARC 2Bs
5-(morpholinomethyl)-3-((5-nitrofurfurylidene)amino)-2-oxazolidinone (dl-form)	139-91-3	IARC 2Bs
5-(morpholinomethyl)-3-((5-nitrofurfurylidene)amino)-2-oxazolidinone (dl-form hydrochloride)	13146-28-6	IARC 2Bs

Chemical Name	CAS#	Hazardous Properties
5-(morpholinomethyl)-3-((5-nitrofurfurylidene)amino)-2-oxazolidinone (l-form)	3795-88-8	IARC 2Bs
Mustard gas	505-60-2	IARC 1; NTPHC
1-Naphthylamine (1-NA)		potential occupational carcinogenic risk
2-Naphthylamine (2-NA)	91-59-8	IARC 1; NTPHC
N-[4-(5-Nitro-2-furyl)-2-thiazolyl]formamide (FANFT)		potential occupational carcinogenic risk
Nitrofuran derivatives		potential occupational carcinogenic risk
Nitrogen mustards		potential occupational carcinogenic risk
2-Nitropropane	79-46-9	IARC 2Bs; NTPAHC
4-Nitroquinoline-1-oxide		potential occupational carcinogenic risk
Nitrosamides		potential occupational carcinogenic risk
Nitrosamines		potential occupational carcinogenic risk
N-Nitroso-N-ethylurea	759-73-9	IARC 2Bs; NTPAHC
N-Nitroso-N-ethylurethane (ENUT)		potential occupational carcinogenic risk
p-Nitrosodiphenylamine	156-10-5	NTPAHC
N-Nitroso-N-methylurea	684-93-5	IARC 2Bs; NTPAHC
N-Nitroso-N-methylurethane (MNUT)		potential occupational carcinogenic risk
N-Nitrosodi-*n*-butylamine	924-16-3	IARC 2Bs; NTPAHC
N-Nitrosodi-*n*-propylamine	621-64-7	IARC 2Bs; NTPAHC
N-Nitrosodiethanolamine	1116-54-7	IARC 2Bs; NTPAHC
N-Nitrosodiethylamine	55-18-5	IARC 2Bs; NTPAHC
N-Nitrosodimethylamine	62-75-9	IARC 2Bs; NTPAHC
N-Nitrosomethylethylamine	10595-95-6	IARC 2Bs
N-Nitrosomorpholine	59-89-2	IARC 2Bs; NTPAHC
N-Nitrosonornicotine	16543-55-8	IARC 2Bs; NTPAHC
Nafenopin	3771-19-5	IARC 2Bs
Naphthalene		irritant
Nickel	7440-02-0	IARC 2A; NTPAHC
Nickel dusts, powders and fumes	7440-02-0	IARC 2A; NTPAHC
Nickel metal/nickel oxide		too hazardous
Nickel carbonate	3333-67-3	IARC 2A; NTPAHC; remove if feasible
Nickel carbonyl	13463-39-3	IARC 2A; NTPAHC
Nickel compounds	7440-02-0	IARC 2A; NTPAHC
Nickel hydroxide [$Ni(OH)_2$]	12054-48-7	IARC 2A
Nickel(II) acetate		suspected carcinogen
Nickel(II) oxide		suspected carcinogen
Nickel oxide	1313-99-1	IARC 2A; NTPAHC
Nickel refining	7440-02-0	IARC 1; NTPHC
Nickel subsulfide	12035-72-2	IARC 2A; NTPAHC
Nickel sulfide fumes and dusts		suspected carcinogen
Nickelocene	1271-28-9	IARC 2A; NTPAHC
Nickelous acetate		remove if feasible
Nicotine		too hazardous; toxic; fire hazard
Niridazole	61-57-4	IARC 2Bs
Nitric acid		corrosive
Nitrogen mustard N-oxide hydrochloride		suspected carcinogen
Norethisterone	68-22-4	IARC 2B; NTPAHC
Octyl phthalate, Di-sec	117-81-7	IARC 2Bs, NTPAHC
Oil shale soot extracts	SEQ NO-24-1	IARC 2Bs
Oral contraceptives, combined	SEQ NO-24-2	IARC 2A
Oral contraceptives, sequential	SEQ NO-24-3	IARC 2B
Osmium tetroxide		too hazardous; toxic; irritant
Oxalic acid		corrosive
Oxygen, tank		too hazardous
Oxymetholone	434-07-1	IARC 2A; NTPAHC
Panfuran S	794-93-4	IARC 2Bs
Paradichlorobenzene		remove if feasible
Paraffin waxes and hydrocarbon waxes, chlorinated	63449-39-8	PAFAR
Paris green		too hazardous; toxic
Pentachloronitrobenzene		suspected carcinogen
Pentachlorophenol (occupational exposure)	87-86-5	IARC 2Bs
Pentane		remove if feasible; fire hazard; explosion hazard
Perchloric acid		irritant; reactive hazard; explosion hazard

Chemical Name	CAS#	Hazardous Properties
Perchloroethylene		suspected carcinogen
Phenazopyridine hydrochloride	136-40-3	NTPAHC
Phenol		suspected carcinogen; too hazardous
Phenoxyacetic acid herbicides (occupational exposure)	SEQ NO-26-0	IARC 2B
Phenoxybenzamine	59-96-1	IARC 2Bs
Phenoxybenzamine hydrochloride	63-92-3	IARC 2Bs; PAFAR
Phenylhydrazine		suspected carcinogen
Phenylthiocarbamide		remove if feasible
1-Phenyl-2-thiourea		remove if feasible; toxic
Phenytoin	57-41-0	IARC 2B; NTPAHC
Phenytoin, sodium salt	630-93-3	NTPAHC
Phosphorus pentoxide		too hazardous; irritant; reactive hazard; fire hazard; corrosive
Phosphorus, red		too hazardous; fire hazard; explosion hazard; reactive hazard
Phosphorus, white		too hazardous; toxic; fire hazard; corrosive
Phthalic anhydride		too hazardous; irritant
Picric acid		too hazardous; toxic; explosion hazard
Polychlorinated biphenyls	1336-36-3	IARC 2B; NTPAHC
Polycyclic aromatic hydrocarbons		potential occupational carcinogenic risk
Ponceau 3R	3564-09-8	IARC 2Bs
Potassium chlorate		remove if feasible; toxic; fire hazard; reactive hazard; explosion hazard
Potassium chromate		suspected carcinogen; remove if feasible; corrosive
Potassium cyanide		toxic; corrosive
Potassium fluoride		corrosive
Potassium hydroxide		corrosive
Potassium oxalate		too hazardous; toxic; corrosive
Potassium periodate		remove if feasible; toxic; irritant; fire hazard
Potassium permanganate		remove if feasible; irritant; fire hazard; reactive hazard
Potassium sulfide		too hazardous; toxic; fire hazard; explosion hazard
Procarbazine (MIH)	671-16-9	IARC 2A; NTPAHC
Procarbazine hydrochloride	366-70-1	NTPAHC
Progesterone	57-83-0	IARC 2B; NTPAHC
Pronamine		suspected carcinogen
Propane sultone		suspected carcinogen
1,3-Propane sultone	1120-71-4	IARC 2Bs; NTPAHC
β-Propiolactone (BPL)	57-57-8	NTPAHC; IARC 2Bs
Pyrogallic acid		too hazardous; toxic
Reserpine	50-55-5	NTPAHC
Rubber industry (certain occupations)	SEQ NO-40-1	IARC 1; NTPHC
Saccharin	81-07-2	NTPAHC; too hazardous
Safrole	94-59-7	IARC 2Bs; NTPAHC
Salol		remove if feasible; toxic
Sodium chromate		too hazardous
Selenium		suspected carcinogen; too hazardous
Selenium sulfide (SeS)	7446-34-6	NTPAHC
Silver compounds		irritant; toxic
Silver cyanide		too hazardous; toxic
Silver nitrate		too hazardous; toxic; irritant
Silver oxide		too hazardous; toxic; fire hazard; explosion hazard
Sodium		irritant; fire hazard; reactive hazard; corrosive; too hazardous
Sodium arsenate	7631-89-2	IARC 1; too hazardous
Sodium arsenite	7784-46-5	IARC 1; too hazardous
Sodium azide		too hazardous; toxic; explosion hazard
Sodium bromate		remove if feasible
Sodium dichromate		suspected carcinogen; too hazardous
Sodium ferrocyanide		too hazardous; corrosive
Sodium fluoride		remove if feasible; toxic

Chemical Name	CAS#	Hazardous Properties
Sodium hydroxide		corrosive
Sodium nitrate		remove if feasible
Sodium nitrite		suspected carcinogen; too hazardous
Sodium oxalate		remove if feasible
Sodium permanganate		toxic; irritant; reactive hazard; fire hazard
Sodium saccharin	128-44-9	IARC 2Bs
Sodium silicofluoride		remove if feasible; toxic; corrosive
Sodium sulfide		too hazardous; toxic; irritant; fire hazard
Sodium thiocyanate		too hazardous; toxic
Soots, tars and mineral oils (occupational exposure)	SEQ NO-29-1	IARC 1
Soots, tars and mineral oils (occupational exposure)	SEQ NO-29-1	NTPHC
Stannic chloride		too hazardous; corrosive; reactive hazard
Stearic acid		suspected carcinogen; too hazardous
Sterigmatocystin	10048-13-2	IARC 2Bs
Streptozotocin	18883-66-4	IARC 2Bs; NTPAHC
Strontium		too hazardous; fire hazard
Strontium chromate	7789-06-2	IARC 1; NTPHC
Sulfallate	95-06-7	IARC 3; NTPAHC
Sulfamethazine		remove if feasible
Sulfonic acid derivatives		potential occupational carcinogenic risk
Sulfuric acid		toxic; irritant; reactive hazard; corrosive
Sulfuric acid, fuming		too hazardous; corrosive
Talc[a]		too hazardous
Tannic acid		suspected carcinogen; too hazardous
Testosterone	58-22-0	IARC 2Bs
Testosterone oenanthate	315-37-7	IARC 2Bs
Testosterone propionate	57-85-2	IARC 2Bs
Tetrabromoethane		too hazardous; toxic; irritant
Tetrachloroethylene	127-18-4	PAFAR
2,3,7,8-Tetrachlorodibenzo-*p*-dioxin	1746-01-6	IARC 2B; NTPAHC
2,3,4,6-Tetrachlorophenol (occupational exposure)	58-90-2	IARC 2Bs
Thermite and compounds		too hazardous; fire hazard
Thioacetamide	62-55-5	IARC 2Bs; NTPAHC; too hazardous
4,4′-thiodianiline	136-65-1	IARC 2Bs
Thiourea	62-56-6	IARC 2Bs; NTPAHC; too hazardous
Thorium dioxide		1314-20-1 NTPHC
Titanium trichloride		too hazardous; fire hazard; irritant
Toluene-2,4-diisocyanate	584-84-9	IARC 3; NTPAHC
Toluene-2,6-diisocyanate	91-08-7	IARC 3
o-Toluidine	95-53-4	IARC 2A; NTPAHC; too hazardous
o-Toluidine hydrochloride	636-21-5	NTPAHC
p-Toluidine		suspected carcinogen
Toxaphene	8001-35-2	IARC 2Bs; NTPAHC
Tremolite	14567-73-8	IARC 1; NTPHC
Treosulphan	299-75-2	IARC 1
Trichloroethylene		suspected carcinogen; remove if feasible
2,2,4-trimethylpentane		toxic; fire hazard
2,4,5-Trichlorophenol (occupational exposure)	95-95-4	IARC 2Bs
2,4,6-Trichlorophenol	88-06-2	IARC 2B; NTPAHC
2,4,6-Trichlorophenol (occupational exposure)	88-06-2	IARC 2Bs
2,4,5-Trichlorophenoxy acetic acid (occupational exposure)	93-76-5	IARC 2Bs
2-(2,4,5-Trichlorophenoxy)propionic acid (occupational exposure)	93-72-1	IARC 2Bs
Trichlorotrifluoroethane		irritant
Tris(1-aziridinyl)phosphate		suspected carcinogen
Tris(2,3-dibromopropyl) phosphate	126-72-7	IARC 2Bs; NTPAHC
Tris(aziridinyl)-para-benzoquinone	68-76-8	IARC 2B
Tris(aziridinyl)-phosphine sulfide	52-24-4	IARC 2B; NTPAHC
Trypan blue		suspected carcinogen
Turpentine		irritant
Uracil mustard	66-75-1	IARC 2B
Uranium		suspected carcinogen; too hazardous

Chemical Name	CAS#	Hazardous Properties
Uranyl acetate		suspected carcinogen; too hazardous
Uranyl nitrate		suspected carcinogen; too hazardous
Urethane		carcinogen; remove if feasible; too hazardous
Vinyl bromide	593-60-2	IARC 3
Vinyl chloride (VC)	75-01-4	IARC 1; NTPHC
Vinyl cyclohexene dioxide		suspected carcinogen
Vinylidene chloride		suspected carcinogen
Vinylite		suspected carcinogen; too hazardous
Wood's metal		too hazardous
Xylene		remove if feasible
Zinc beryllium silicate	39413-47-3	IARC 2A; NTPAHC
Zinc chromate	13530-65-9	IARC 1; NTPHC
Zinc chromate hydroxide	15930-94-6	IARC 1
Zinc chromates		suspected carcinogen

[a]Suggested alternative:
For methyl orange and methyl red, bromophenol blue and bromothymol blue.
For talc, starch talc.

IARC Categories

Group	Hazardous Properties
1	causally associated with human cancer
2A	probably carcinogenic to humans; higher degree of evidence
2B	probably carcinogenic to humans; lower degree of evidence
2Bs	probably carcinogenic to humans; lower degree of evidence; evaluated subsequent to IARC Supplement 4
2B*	IARC animal carcinogens for which human data are not available; considered by OSHA to correspond to Group 2B
3	IARC animal carcinogens for which human data are not available; to be safe, these should be treated as carcinogens

NTP Human Carcinogens

NTPHC = National Toxicology Program Human Carcinogens
NTPAHC = National Toxicology Program Anticipated Human Carcinogens
PAFAR = Proposed for Addition in NTP's Fifth Annual Report on Carcinogens

APPENDIX E: ADDRESSES AND TELEPHONE NUMBERS FOR NIOSH, OSHA, AND EPA OFFICES AROUND THE COUNTRY

THE NATIONAL INSTITUTE FOR OCCUPATIONAL SAFETY AND HEALTH (NIOSH)

Technical Information and Assistance
1-800-35-NIOSH

Publications
513-533-8287

Office of the Director
1600 Clifton Road
Atlanta, GA 30333
404-639-3061

Division of Safety Research
944 Chestnut Ridge Road
Morgantown, WV 26505-2888
304-291-4595

Washington Office
200 Independence Ave., SW
Washington, DC 20201
202-472-7134

Division of Biomedical and Behavioral Science
4676 Columbia Parkway
Cincinnati, OH 45226-1998
513-533-8465

Division of Physical Science and Engineering
4676 Columbia Parkway
Cincinnati, OH 45226-1998
513-841-4321

Division of Training and Manpower Development
4676 Columbia Parkway
Cincinnati, OH 45226-1998
513-533-8221

Division of Standards Development and
Technology Transfer
4676 Columbia Parkway
Cincinnati, OH 45226-1998
513-533-8302

Division of Surveillance, Hazard Evaluations and
Field Studies
4676 Columbia Parkway
Cincinnati, OH 45226-1998
513-841-4428

Regional Offices

NIOSH
Room 1401
JFK Federal Building
Boston, MA 02203
617-565-1443

NIOSH
Suite 1110
101 Marietta Tower
Atlanta, GA 30323
404-331-2396

NIOSH
1185 Federal Building
1961 Stout Street
Denver, CO 80294
303-844-6166

OCCUPATIONAL SAFETY AND HEALTH ADMINISTRATION (OSHA)

Headquarters Office

U.S. Department of Labor
Occupational Safety and Health Administration
3rd and Constitution Avenue, N.W.
Washington, D.C. 20210
202-523-8148

Publications
202-523-9667

Regional Offices

Region I (Maine, Vermont, New Hampshire, Massachusetts, Rhode Island, Connecticut)

16-18 North Street
1 Dock Square Bldg., 4th Floor
Boston, MA 02109
617-565-1145

Region II (New York, New Jersey, Puerto Rico)

201 Varick Street
New York, NY 10014
212-337-2378

Region III (Pennsylvania, Delaware, Maryland, West Virginia, Virginia, District of Columbia)

Gateway Bldg., Suite 2100
3535 Market Street
Philadelphia, PA 19104
215-596-1201

Region IV (Kentucky, Tennessee, North Carolina, South Carolina, Georgia, Alabama, Mississippi, Florida)

1375 Peachtree Street, N.E.
Suite 587
Atlanta, GA 30367
404-347-3573

Region V (Ohio, Michigan, Indiana, Illinois, Wisconsin, Minnesota)

230 South Dearborn Street
32nd Floor, Room 3244
Chicago, IL 60604
312-353-2220

Region VI (Louisiana, Arkansas, Texas, Oklahoma, New Mexico)

525 Griffin Square, Room 602
Dallas, TX 75202
214-767-4731

Region VII (Missouri, Iowa, Kansas, Nebraska)

911 Walnut Street, Room 406
Kansas City, MO 64106
816-374-5861

Region VIII (North Dakota, South Dakota, Colorado, Wyoming, Montana, Utah)

Federal Bldg., Room 1576
1961 Stout Street
Denver, CO 80294
303-844-3061

Region IX (Arizona, Nevada, California, Hawaii)

71 Stevenson Street, 4th Floor
San Francisco, CA 94105
415-995-5672

Region X (Idaho, Oregon, Washington, Alaska)

Federal Office Bldg., Room 6003
909 First Avenue
Seattle, WA 98174
206-442-5930

U.S. ENVIRONMENTAL PROTECTION AGENCY (EPA) REGIONAL OFFICES

Region 1

John F. Kennedy Federal Building
Boston, MA 02203
617-223-7210

Region 2

26 Federal Plaza
New York, NY 10278
212-264-2525

Region 3

Curtis Building
6th and Walnut Street
Philadelphia, PA 19106
215-597-9814

Region 4

345 Courtland Street, NE
Atlanta, GA 30365
404-881-4727

Region 5

230 South Dearborn Street
Chicago, IL 60604
312-353-2000

Region 6

1201 Elm Street
Dallas, TX 75270
214-767-2600

Region 7

324 East 11th Street
Kansas City, MO 64106
816-926-3720

Region 8

1860 Lincoln Street
Denver, CO 80295
303-837-3895

Region 9

215 Fremont Street
San Francisco, CA 94105
415-974-8153

Region 10

1200 Sixth Avenue
Seattle, WA 98101
206-442-5810

INDEX

INDEX

Italicized numbers denote pages on which definitions or descriptions appear.